# GREEN
## BUILDING
## RESOURCE
# GUIDE

# GREEN BUILDING RESOURCE GUIDE

JOHN HERMANNSSON AIA
ARCHITECT

The Taunton Press

Publisher: Jon Miller
Acquisitions Editor: Julie Trelstad
Editorial Assistant: Karen Liljedahl

Copy/Production Editor: Diane Sinitsky
Designer/Layout Artist: Lynne Phillips
Cover Drawing: Bobbi Angell

Typeface: Stone Serif, Giltus
Paper: 70-lb. Springhill Recycled Offset
Printer: Bawden Printing, Eldridge, Iowa

**Taunton**
BOOKS & VIDEOS
*for fellow enthusiasts*

First printing: 1997
Printed in the United States of America

A FINE HOMEBUILDING Book
FINE HOMEBUILDING® is a trademark of The Taunton Press, Inc.,
registered in the U.S. Patent and Trademark Office.

The Taunton Press, 63 South Main Street, PO Box 5506,
Newtown, CT 06470-5506
e-mail: tp@taunton.com

**Library of Congress Cataloging-in-Publication Data**
Hermannsson, John.
    Green building resource guide / John Hermannsson.
        p.    cm.
    Includes biographical references and index.
    ISBN 1-56158-219-0
        1. Building materials—United States—Catalogs.    2. Green
products—United States—Catalogs.    I. Title.
    TH455.H47    1997
    691' .029'473—dc21                    96-51637
                                    CIP

To those who will become the pioneers
of a sustainable design.

# CONTENTS — Listed by Construction Division

# PREFACE

As an architect, I have been driven by client requests for environmentally conscious designs to create this guide. My ability to satisfy these client needs has been frustrated by a lack of easy access to green building product information. Now, this guide provides that information in an accessible, organized way.

In addition, this guide addresses one of the first concerns of clients: cost. "How much will substituting a green building product for a conventional product cost?" This is a reasonable question, and there should be a reasonable way to provide an answer. A client may get a typical response from a builder such as, "Any deviation from standard construction practice will probably increase the cost of your project." My experience has been that such a statement is usually enough to kill a good idea. Most clients would simply set aside any thoughts of green building in favor of the known costs of conventional construction. This has led to lost opportunities for clients because, in fact, many green building products and materials are about the same cost as conventional items.

The *Green Building Resource Guide* features a unique tool for addressing this problem of green building costs. A Price Index Number assigned to each listing compares the cost of a green product to a commonly used product that it could replace. With this cost information, now shopping around for the best buy is at your fingertips. With the Price Index Number, this guide sets a new standard for green building catalogs.

You may notice this guide is different from many other green building publications because it doesn't have illustrations, testimonials, or inspirational essays. It just has information accompanied by a well-organized system of indices. It assumes you're already inspired and want to work with green products. Moreover, it assumes you want access to green building information to be quick and efficient, like finding the right tool in an orderly workshop. This isn't a book for the coffee table; it's a green building tool for making it all happen, so architects will specify, builders will build, and clients will realize more value with green building.

The main focus of the *Green Building Resource Guide* is standard "unheroic" residential construction products and materials, such as framing, siding, and flooring materials, rather than the higher-profile "heroic" products, such as solar or photovoltaic panels. "Unheroic" materials consume most of the energy and resources of the residential building industry. They are the materials most architects, builders, and homeowners are most familiar with. Substituting many conventional materials with green building materials listed in this guide would change the course of residential construction. A mutually beneficial situation would be created by providing a market for the waste stream, thereby improving the economy and preserving our natural resources and environment for future generations. Nontoxic green building materials would improve air quality and health, particularly for chemically hypersensitive people, whose numbers are increasing.

Although the creation of the products and materials in this guide was inspired by resource conservation and health, the survival of this burgeoning manufacturing effort is still dependent upon a market for the goods and our willingness to build with green products. But, then again, isn't the survival of the green industry today really our survival tomorrow? In the final analysis, how can we afford not to use these products? To build for our future means to build it well, to build it beautifully, and to build it green.

# INTRODUCTION

I'm sure you will appreciate the elegant simplicity of so many of the green building products appearing in this guide. Consider for a moment a beautiful floor with a soft luster, an inviting visual depth, and a hard granite-like finish that is not cold to the touch. It is made of compressed soy beans. The Minnesota factory of Phenix Biocomposites can be mistaken for a bakery because of the pleasant aroma it emits while roasting its soy bean ingredients.

If you've ever cut particleboard with a power saw, you may have experienced the unpleasant effects of formaldehyde outgassing from the resin binders. Try cutting formaldehyde-free WheatBoard. You may imagine you're slicing a fresh loaf of wheat bread! PrimeBoard is the North Dakota company that developed this product as a substitute for particle-board. They have also created a market for farmers to sell their straw for recycling. Like this one, really good ideas are usually good in more ways than one.

The Arab Oil Embargo of 1973 mobilized an extraordinary campaign to conserve thermal energy resources in the heating and cooling of buildings. We are now the beneficiaries of these great advances made over the last two decades. Most states have incorporated performance or prescriptive thermal energy requirements into their building codes. As a shining example, in California alone, the state claims that the energy efficiency building regulations in effect since 1978 have saved $11.4 billion from "going up the flue." The California Energy Commission forecasts that an additional $43 billion in utility bills will be saved by the year 2011.

Residential operating costs for energy-efficient houses have been reduced to a point of diminishing returns. A passive solar house built today can be designed to consume less energy over the next several decades than was consumed to build it. If comparable gains in energy conservation are to be realized over the next 20 years, embodied energy must be the focus.

Embodied energy is essentially all of the other energy consumed in the life cycle of a building: harvesting resources, manufacturing products, transporting materials, building, and recycling or removing structures.

Moreover, thermal energy efficiency has a side effect: It may be hazardous to your health. Toxic outgassing of many synthetic materials used in construction can be trapped within a house much longer now because air infiltration has been greatly reduced in today's residences over the houses of 20 years ago. Low air infiltration is an important feature of thermal energy efficiency. Using nontoxic building materials for healthier indoor air quality is even more essential today.

Ultimately, homeowners, architects, and builders must aspire to a sustainable design. Sustainable design means solving the needs of the present without detracting from the needs of the future by creating architecture that minimizes the use of natural resources, toxic materials, and emissions of waste and pollutants over the life cycle of a building. The materials used in a sustainably designed structure are green building materials. The use of green building materials is a responsible approach to the maintenance of human health, the conservation of nonrenewable resources, and the preservation of the environment for future generations.

Although not all products in this guide are necessarily as "green" as the examples I have given here, they do possess at least one of the following characteristics: nontoxicity (for indoor air quality), recycled content, or durability (requiring little, if any, maintenance). Many of these products are evolving in an emerging green building industry restless with invention, alive with opportunity, and fueled by a passionate belief in our sustainable future.

Here's to our future.

# HOW TO USE THIS BOOK

The *Green Building Resource Guide* provides three ways to access information. First, the guide is organized by construction division section numbers, so a design or building professional may open directly to familiar section numbers for specific products. The construction division section numbers appearing in the guide are listed on the table of contents.

Since section number names are not always clear to someone unfamiliar with the system, a product type index (see pp. 141-152) provides more descriptive subcategories. A typical homeowner may prefer to use this index first.

In addition, a manufacturer index (see pp. 129-140) assists people in quickly refinding a previously located listing, since most people usually remember the manufacturer associated with a product rather than a product category.

# PRICE INDEX NUMBER & PRICE INDEX STANDARD

The Price Index Number is based on the assumption that although labor and material costs can vary from one location to another in the United States, the relative costs of similar products to one another is a constant number and remains more stable over time than that of price. This assumption is the basis for construction cost-estimating books that use local multipliers to adjust cost estimates for different locations. Consequently, the Price Index Number is consistent with cost information available in such books as the *Dodge Repair & Remodel Cost Book* and is a convenient tool for rapidly estimating residential construction costs.

The Price Index Number was obtained by asking manufacturers to compare the costs of their green products to commonly available conventional products. These conventional products, which could be replaced by particular green products, are the Price Index Standards.

The costs associated with each Price Index Standard were obtained from distributors and retailers and were verified with a construction cost book.

# GUIDE TO ICON SYMBOLS

 **Nontoxic:** Manufacturer claims product or material is not poisonous when used as intended.

 **Recycled content:** Products manufactured from used products or by-products. When available, the percentage of recycled content is included in the product description.

 **Resource efficient:** Products or materials that are more conserving of energy or materials than similar conventional products.

 **Long life cycle:** Products or materials that last significantly longer and consume less resources for maintenance than conventional products or materials.

 **Environmentally conscious:** Products or materials that are manufactured or provided in a way that reduces a negative environmental impact compared to similar conventional products or materials.

## 1 Bestmann Green Systems, Inc.

53 Mason St., Salem, MA 01970
Phone: 508 741 1166    Fax: 508 741 3780

**Price Index Number:**    **0.60 (typical project)**
**Price Index Standard:**    **geotextile and concrete**

### SOIL STABILIZATION

Bioengineering approach to utilizing nature to stabilize steep embankments and to reduce erosion along riverbanks and ocean beaches. Provides an earth-friendly alternative to typical retaining walls by using biodegradable coconut-fiber mats and rolls and plant plugs. Company also offers a nationwide consulting service. A subsidiary company of Bestmann of Germany with 20 years of pioneering experience in ecological solutions to erosion control in Europe.

## 2 Bonded Fiber Products, Inc.

2748 Tanager Ave., Commerce, CA 90040
Phone: 213 726 7820    Fax: 213 726 2805

**Price Index Number:**    **1.0**
**Price Index Standard:**    **geotextile**

### SOIL STABILIZATION

Geotextiles for erosion control from 90% recycled PET plastics.

## 3 Environmental Plastics, Inc.

4981 Keelson Dr., Columbus, OH 43232
Phone: 614 861 2107    Fax: 614 445 6907

**Price Index Number:**    **0.90**
**Price Index Standard:**    **concrete retaining wall + labor**

### RETAINING WALL SYSTEM

C-LOC corrugated plastic retaining wall, noise wall, and seawall made from 86% recycled PVC plastics. Meets federal specifications. Available in UV-deflecting colors of gray, white, light yellow, light green, and light blue. Graffiti-resistant and cleans easily. Does not require heavy equipment for installation. Wall can be installed by two workers with one jackhammer.

## 4 Presto Products

PO Box 2399, Appleton, WI 54913
Phone: 800 548 3424 or 414 739 9471
Fax: 414 738 1418

**Price Index Number:**    **0.5-0.75**
**Price Index Standard:**    **concrete retaining wall + labor**

### RETAINING WALL SYSTEM

GEOWEB cellular confinement system is an engineered HDPE, expanded honeycomb-like matrix infilled with earth to provide stable earth retention structures with variable slope that reinforce roots against erosion. Computerized analysis is available to optimize GEOWEB designs and to estimate costs. Also available with perforated cell walls for drainage and in various colored faces.

## 5 Amazing Recycled Products

PO Box 312, Denver, CO 80201
Phone: 800 241 2174 or 303 699 7693
Fax: 303 699 2102

**Price Index Number:** 0.30
**Price Index Standard:** asphalt speed bump

### SITE TRAFFIC CONTROL
Traffic control products manufactured from recycled rubber, including removable speed bumps 48 in. by 12 in. by 2¼ in. high.

## 6 AMREC

130 Sturbridge, Charlton, MA 01507
Phone: 508 248 3777    Fax: 508 248 4911

**Price Index Number:** 0.33-0.50
**Price Index Standard:** hot mix asphalt

### ASPHALTIC CONCRETE PAVING
Cold mix asphalt manufactured from 100% recycled petroleum-contaminated soil and recycled asphalt shingles. Stored as an asphaltic emulsion with a shelf life of about 6 months. Requires higher ambient temperature, dry weather, and longer curing time than hot mix asphalt.

## 7 Cunningham Brick Co., Inc.

701 N. Main St., Lexington, NC 27292
Phone: 800 672 6181 or 910 248 8541 (plant)
Fax: 910 224 0002 or 910 472 6181 (plant)

**Price Index Number:** 3.0
**Price Index Standard:** gravel

### RECYCLED BRICK GRAVEL
BRICK NUGGETS are manufactured from reprocessed oil-contaminated soil as small, medium, and large pebbles and rocks for walks and drives. The largest average 1 in. to 3¾ in. in diameter, are jagged in shape, and are red colored. Distribution is typically limited to Eastern Canada down to Florida and west to Texas.

## 8 Eagle One Golf Products

1201 W. Katella Ave., Orange, CA 92867
Phone: 800 448 4409 or 714 997 1400
Fax: 714 997 3400

**Price Index Number:** 0.80
**Price Index Standard:** virgin rubber mats

### SYNTHETIC SURFACING
100% recycled car-tire rubber for interior or exterior pathways and stairs with a nonslip surface. Product is available as a 4-ft. by 6-ft. by ¾-in. mat with ridges that can be glued down with a construction adhesive.

## 9 International Surfacing, Inc.

6751 W. Galveston, Chandler, AZ 85226
Phone: 800 528 4548, 800 829 1144 (AZ only), or 602 268 0874
Fax: 602 961 0766

**Price Index Number:** 1.4
**Price Index Standard:** asphalt surfacing

### ASPHALTIC CONCRETE PAVING
Asphalt rubber surfacing is a composite material manufactured from recycled tires and asphalt. Double the strength of standard asphalt, this surfacing requires only one-half the material and thickness of standard asphalt for a typical application. Also, it lasts longer and has improved skid-resistant properties over asphalt. Currently only available for large-scale roadway applications.

## 10  M. Susi and Sons

21 Westwood, Dorchester, MA 02121
Phone: 617 265 4525   Fax: 617 265 5229

**Price Index Number:   0.90**
**Price Index Standard:   gravel**

### RECYCLED CONCRETE GRAVEL

Recycled concrete is crushed into a screened-gravel product for use as a subbase for road work and foundation drainage. Compacts and drains better than gravel. Distribution is limited to the greater Boston area and Newton, Mass.

## 11  RB Rubber Products, Inc.

904 E. 10th, McMinnville, OR 97128
Phone: 800 525 5530 or 503 472 4691
Fax: 503 434 4455

**Price Index Number:   4.0 (payback in 4 yrs)**
**Price Index Standard:   redwood chips**

### SYNTHETIC SURFACING

BOUNCEBACK loose-fill rubber chips are used as an alternative to redwood chips for playground surfacing. The rubber chips are about thumb size and are suitable for handicap access. All rubber products are made from 100% recycled tire rubber.

## 12  Safety Turf

PO Box 820, Oaks, PA 19456
Phone: 610 666 1779 or 800 854 4595, (PA only)
Fax: 610 666 1768

**Price Index Number:   3.7-6.7**
**Price Index Standard:   asphalt surfacing**

### SYNTHETIC SURFACING

Recycled rubber playground surfacing applied in two layers: 1½-in. base from recycled tire rubber with a ⅜-in. finish surface of EPDM rubber in various colors. Business limited to East Coast.

## 13  Soil Stabilization Products Co.

PO Box 2779, Merced, CA 95344
Phone: 800 523 9992 or 209 383 3296
Fax: 209 383 7849

**Price Index Number:   0.05-0.14**
**Price Index Standard:   asphalt surfacing**

### ROAD SURFACING

EMC SQUARED is a biocatalyst treatment for improving the stability of compacted earth. The biocatalyst is applied as a liquid mixed with water in various concentrations from 1 part per 30 to 1 part per 600, depending upon the application. Applications include soil stabilization of unsurfaced roads and road shoulders, slopes, and embankments against wind and water erosion. Manufacturer claims EMC SQUARED is nonflammable, nonhazardous, and acceptable for environmentally sensitive areas.

## 14  Surfacing Concepts, Inc.

25875 Jefferson St., Clair Shores, MI 48081
Phone: 810 776 5560   Fax: 810 463 7597

**Price Index Number:   1.25, TIRECRETE**
**Price Index Standard:   asphalt surfacing**

### RUBBER SURFACING

TIRECRETE is a wire-mesh reinforced porous exterior rubber surfacing made from 50% recycled tires, for roadways and driveways. CUSHION AIR is a 3-in.-thick porous rubber surface primarily for playgrounds, made from recycled tires.

## 15  Cherokee Sanford Brick Co.

1600 Colon Rd., Sanford, NC 27330
Phone: 800 277 2700   Fax: 919 774 5300

**Price Index Number:    1.0**
**Price Index Standard:  brick**

**BRICK PAVERS**
Brick pavers manufactured from reprocessed oil-contaminated soil. Ideal for use in walkways, driveways, and patios. Comes in a standard brick of 8 in. by 4 in. by 2¼ in. or modular brick of 7⅝ in. by 3⅝ in. 2¼.

## 16  Playfield International

PO Box 8, Chatsworth, GA 30705
Phone: 800 685 7529 or 706 695 4581
Fax: 706 695 4755

**Price Index Number:    4.7**
**Price Index Standard:  brick**

**RUBBER PAVERS**
I-BLOCK PAVERS are interlocking ½-in.-thick 100% recycled rubber pavers for walkways and patios. Other products include troweled-in-place (porous), cast-in-place (nonporous), and interior rubber sheet flooring manufactured from 100% recycled rubber with a urethane binder. Installations include skating rinks and walking areas.

## 17  Bomanite Corp.

PO Box 599, Madera, CA 93639-0599
Phone: 800 854 2094 or 209 673 2411
Fax: 209 673 8246

**Price Index Number:    4.0**
**Price Index Standard:  asphalt surfacing**
                         **with drains**

**POROUS PAVING**
GRASCRETE is a porous-paving system designed for heavy-duty vehicle use. Bomanite manufactures the fiberglass forms used by certified applicators to create a reinforced poured-in-place monolithic concrete grid through which grass can grow. Applicable to driveways and parking lots.

## 18  Invisible Structures, Inc.

14704 D E. 33rd Pl., Aurora, CO 80011-1218
Phone: 800 233 1510 or 303 373 1234
Fax: 303 373 1223

**Price Index Number:    0.85**
**Price Index Standard:  asphalt surfacing**
                         **with drains**

**POROUS PAVING**
GRASSPAVER2 manufactured from 100% recycled HDPE for porous paving applications supporting up to 5700 psi. Installations include the Orange Bowl parking lot in Miami, Fla. and the Presidio in San Francisco, Calif.

## 19  Presto Products

PO Box 2399, Appleton, WI 54913
Phone: 800 548 3424   Fax: 414 738 1418

**Price Index Number:    2.0-2.7**
**Price Index Standard:  asphalt surfacing**

**POROUS PAVING**
GEOBLOCK porous-pavement system is a series of gridded, interlocking, high-strength plastic paving blocks. Suitable for turf protection in load-bearing applications, such as utility access lanes, parking areas, and driveways. Grass will grow through the block and completely cover it in about 7 months. Maximum allowable grade is 10%. Made from a minimum of 50% post-consumer recycled plastic.

## 20 Global Technology Systems

PO Box 25, 26112 110th St., Trevor, WI 53179
Phone: 800 558 3206 or 414 862 2311
Fax: 414 862 2500

**Price Index Number:** 2.0/1.0
**Price Index Standard:** concrete/clay tennis court

### RUBBER SURFACING

Composite flooring manufactured from 90% recycled rubber with a binder used on tennis courts, pool decks, and running surfaces. This product is a base material available in rolls 48 in. by 30 ft. or 60 ft. in the form of tiles 37 in. by 37 in. by 3mm to ½ in. thick. Available in four densities from 40 lb./cu. ft. to 60 lb./cu. ft. Also manufacturer of cork tiles made from recycled cork from the wine industry.

## 21 Playfield International

PO Box 8, Chatsworth, GA 30705
Phone: 800 685 7529 or 706 695 4581
Fax: 706 695 4755

**Price Index Number:** 0.71
**Price Index Standard:** Astroturf

### RUBBER SURFACING

PLAYFIELD TILES come in 1m by 0.5m by 1½ in., 2 in., or 3 in. thicknesses. Made from 100% recycled rubber. Easily installed as a one-component system. Has a nonskid surface and is used on McDonald's restaurants' playgrounds. Other products include troweled-in-place (porous), cast-in-place (nonporous), and interior rubber sheet flooring manufactured from 100% recycled rubber with a urethane binder. Installations include skating rinks and walking areas.

## 22 EEE ZZZ Lay Drain Co., Inc.

PO Box 867, Pisgah Forest, NC 28768
Phone: 800 649 0253 or 704 883 2130
Fax: 704 884 2348

**Price Index Number:** 0.25
**Price Index Standard:** gravel + labor

### GRAVEL SUBSTITUTE

Gravel substitute for use in foundations and interceptor drainage, septic system absorption fields, and various ground-water control systems. Product is manufactured from 65% recycled EPS waste from primarily industrial and commercial sources.

## 23 Plastic Tubing, Inc.

PO Box 878, Rosoboro, NC 28382
Phone: 800 334 6602 or 910 525 5121
Fax: 910 525 4934

**Price Index Number:** 1.0
**Price Index Standard:** virgin plastic tubing

### PLASTIC TUBING

CORR-A-FLEX is a plastic flexible tubing for site drainage that comes in slotted, solid, or double walled designs, in 10-ft. or 20-ft. lengths, or 4-in. to 15-in. coils. Tubing is manufactured from up to 70% recycled HDPE from post-industrial and post-consumer sources. High recycled content may be specified in order.

### 24 Resource Conservation Tech.

2633 N. Calvert St., Baltimore, MD 21218
Phone: 800 477 7724 or 410 366 1146
Fax: 410 366 1202

**Price Index Number:** 1.0-3.0
**Price Index Standard:** standard rubber

**POND LINER**
Distributor of butyl and EPDM pond liners. 30-mil butyl
pond liners are manufactured talc-free and are fish safe. Butyl
rubber remains flexible in cold weather, conforms to irregular
shapes, and lasts almost indefinitely, according to the
manufacturer. 45-mil EPDM pond liners also last much longer
than other premium liners and are as flexible as plastic liners
half the thickness.

### 25 Mandish Research International

5055 State Rd. 46, Mims, FL 32754
Phone: 407 267 2561   Fax: 407 268 1972

**Price Index Number:** 0.10-0.20
**Price Index Standard:** carved stone
ornament

**LANDSCAPE ORNAMENTS**
Designer and manufacturer of molds for making precast,
lightweight Donolite concrete products manufactured from a
mixture of cement and recycled fiberglass, polystyrene, and
tire rubber. Manufacturer offers 5,000 molds for various
applications including birdbaths, fence posts, landscape
timbers, architectural balustrades, and various landscape
ornaments. Molds can be purchased from Mandish.

### 26 Schuyler Rubber Co.

16901 Wood-Red Rd., Woodinville, WA 98072
Phone: 800 426 3917 or 206 488 2255
Fax: 206 488 2424

**Price Index Number:** 0.50-0.90
**Price Index Standard:** extruded virgin
rubber

**WALKWAY AND ROADWAY ADDITIONS**
Recycled tire products including dock bumpers, wheel
chocks, truck bumpers, and corner guards.

### 27 Scientific Developments, Inc.

PO Box 2522, Eugene, OR 97402
Phone: 800 824 6853 or 541 686 9844
Fax: 541 485 8990

**Price Index Number:** 0.50-0.90/0.24-0.33
**Price Index Standard:** extruded virgin
rubber/asphalt bump

**WALKWAY AND ROADWAY ADDITIONS**
Full line of recycled tire products including dock bumpers,
wheel chocks, speed bumps, trailer bumpers, delineators, and
delineator bases.

### 28 AquaPore Moisture Systems

610 S. 80th Ave., Phoenix, AZ 85043
Phone: 800 635 8379 or 602 936 8083
Fax: 602 936 9040

**Price Index Number:** 1.0
**Price Index Standard:** virgin rubber soaker
hose or perforated
PVC pipe

**DRIP-IRRIGATION HOSE**
MOISTURE MASTER drip-irrigation hoses manufactured from
recycled tire rubber for aboveground or underground
irrigation of gardens and landscapes. Hose sizes are ½ in. and
⅝ in. diameters in lengths of 25 ft., 50 ft., and 75 ft. A
controlled amount of water can be distributed through these
hoses, which are porous throughout, providing an efficient
use of water resources.

## 29 All Fiberglass Products Corp.

102 S. Midland, Rockbale, IL 60436
Phone: 800 438 7395 or 815 729 3326
Fax: 815 741 0058

**Price Index Number:** n/a
**Price Index Standard:** n/a

### FENCE SCREENING

Fiberglass strips manufactured from industrial waste from trailer linings in the food-processing industry. Horizontally applied to chain-link fences to screen the view. Long-lasting riveted together with aluminum or nylon, and can be manufactured in various colors. Typical applications include tennis courts, garden centers, and industrial sites.

## 30 Hygrade Glove and Safety

30 Warsoss Pl., Brooklyn, NY 11205
Phone: 800 237 8100  Fax: 718 694 9500

**Price Index Number:** 1.0
**Price Index Standard:** fence from new material

### FENCES

Safety fence for construction sites made from 100% recycled polyethylene.

## 31 Children's Playstructures, Inc.

12441 Mead Way, Littleton, CO 80125
Phone: 800 874 9943 or 303 791 7626
Fax: 303 874 9943

**Price Index Number:** 0.75
**Price Index Standard:** equipment without recycled content

### PLAYGROUND EQUIPMENT

Commercial playground equipment manufactured from 100% recycled HDPE. Delivery and installation included in Colorado

## 32 Florida Playground and Steel Co.

4701 S. 50th St., Tampa, FL 33619-9500
Phone: 800 444 2655  Fax: 813 247 1068

**Price Index Number:** 1.6
**Price Index Standard:** pine site furniture

### PLAYGROUND EQUIPMENT

Site furniture, such as benches, tables, and trash receptacles, manufactured from recycled plastics. The company also makes car stops. Playground equipment is also made mostly of steel with recycled plastic components.

## 33 BTW Industries, Inc.

3939 Hollywood Blvd., Hollywood, FL 33021
Phone: 954 962 2100  Fax: 954 963 4778

**Price Index Number:** 2.0
**Price Index Standard:** pressure-treated southern pine

### SITE FURNISHINGS

ENVIROWOOD lumber in all common nominal sizes, benches, picnic tables, car stops, decking, and railings made from 100% recycled commingled post-consumer plastic.

## 34 Cambridge Designs

Rt. 4, Box 188, Greenville, AL 36037
Phone: 800 477 7320   Fax: 334 382 9207

**Price Index Number:   0.95**
**Price Index Standard:   concrete furniture**

**SITE FURNISHINGS**
Site furniture from 100% recycled HDPE plastics. Products include picnic tables, benches, trash receptacles, and water fountains.

## 35 Eagle One Golf Products

1201 West Katella Ave., Orange CA 92867
Phone: 800 448 4409 or 714 997 1400
Fax: 714 997 3400

**Price Index Number:   1.0**
**Price Index Standard:   wood benches**

**SITE FURNISHINGS**
100% recycled plastic benches, posts, and corral fencing.

## 36 Earth Care Midwest

PO Box 536, Lake Odessa, MI 48849
Phone: 888 753 2784 or 616 374 7443
Fax: 616 374 0907

**Price Index Number:   2.0**
**Price Index Standard:   wood products**

**SITE FURNISHINGS**
DURATECH PRODUCTS include benches, tables, bike racks, trash receptacles, and lumber manufactured from 100% recycled post-consumer HDPE plastic lumber. 15-year guarantee.

## 37 Earth Safe

PO Box 1401, Marstons Mills, MA 02648-0014
Phone: 508 420 5681   Fax: 508 420 4214

**Price Index Number:   2.5**
**Price Index Standard:   pressure-treated
                          lumber**

**SITE FURNISHINGS**
Fabricator of TREX composite lumber products, such as benches, tables, posts, and fencing, made from recycled plastic LDPE and wood fiber. Also fabricators of TRIMAX products made from 80% HDPE and 20% fiberglass. TRIMAX works well in marine environments. 10-year guarantee for TREX and 50 year guarantee for TRIMAX.

## 38 EnviroSafe Products, Inc.

81 Winant Pl., Staten Island, NY 10309-1311
Phone: 718 984 7272   Fax: 718 984 1083

**Price Index Number:   1.0-1.5**
**Price Index Standard:   select heart redwood**

**SITE FURNISHINGS**
Recycled plastic lumber products, including benches and tables.

### 39 Giati Design

614 Santa Barbara St., Santa Barbara, CA 93101
Phone: 805 965 6535   Fax: 805 965 6295

**Price Index Number:** 2.0
**Price Index Standard:** wood furniture

**SITE FURNISHINGS**
Manufacturer of teak benches, tables, and chairs for exterior use. All teak comes from sustainably harvested sources in Indonesia and southern Asia certified by the Rainforest Alliance. Manufacturer claims high-quality product.

### 40 The Plastic Lumber Co.

540 S. Main St., Bldg. 7, Akron, OH 44311
Phone: 800 886 8990 or 330 762 8989
Fax: 330 762 1613

**Price Index Number:** 1.1-1.2
**Price Index Standard:** wood site furnishings

**SITE FURNISHINGS**
Site furnishings manufactured from recycled milk jugs and HDPE plastics. Products include picnic tables, trash bins, benches, signage, and planters.

### 41 Recreation Creations

215 W. Mechanic St., Hillsdale, MI 49242
Phone: 800 766 9458   Fax: 517 439 1878

**Price Index Number:** 1.3
**Price Index Standard:** redwood furniture

**SITE FURNISHINGS**
Site furniture and playground equipment from recycled plastics and recycled steel.

### 42 Smith and Hawkins

117 E. Strawberry Dr., Mill Valley, CA 94941
Phone: 800 776 3336 or 415 389 8300
Fax: 415 383 7243

**Price Index Number:** 1.0
**Price Index Standard:** wood furniture

**SITE FURNISHINGS**
Manufacturer of teak benches, tables, and chairs for exterior use. All teak comes from sustainably harvested sources in Java and has been awarded Smart Wood certification by the Rainforest Alliance.

### 43 Davis Colors

3700 E. Olympic Blvd., Los Angeles, CA 90023
Phone: 800 356 4848 or 213 269 7311
Fax: 213 269 2867

**Price Index Number:** 0.60
**Price Index Standard:** concrete sealers
with VOCs

**CONCRETE COLORANT AND SEALERS**
Manufacturer of inorganic, mineral-based inert pigments for integral coloring of finished concrete and water-based clear and pigmented sealers for concrete. Meets all California environmental standards. Environmentally safe earth tones, reds, and yellows made from natural and synthetic iron oxides.

## 44   Englehard Corp.

220 W. Westfield Ave., Roselle Park, NJ 07204
Phone: 908 245 9500   Fax: 908 245 6469

**Price Index Number:**   **1.0**
**Price Index Standard:**   **standard foaming agents**

### CONCRETE FOAMING LIQUID

MEARLCRETE foaming liquid is for creating cellular concrete or foamed concrete by either being added to cement on-site or at a precast manufacturing facility. Mainly used for reducing the weight of concrete while maintaining load-bearing capacity. Can produce insulating cellular concrete weighing from 20 to 120 lb./cu. ft. The foaming agent is protein based but proprietary, designed to be nonhazardous and nonpolluting.

## 45   L. M. Scofield Co.

6533 Bandini Blvd., Los Angeles, CA 90040-3182
Phone: 800 800 9900 or 213 720 3000
Fax: 213 720 3030

**Price Index Number:**   **0.77-0.85**
**Price Index Standard:**   **clear concrete sealers with VOCs**

### CONCRETE COLORANT AND SEALERS

Manufacturer of water-based clear and pigmented sealers for concrete. Also produces water-soluble integral coloring admixture for finished concrete and REPELLO sacrificial surface coating to guard against graffiti. Graffiti can then be removed with a high-power washer. Meets all California environmental standards.

## 46   3-10 Insulated Forms LP

PO Box 460790, Omaha, NE 68046
Phone: 800 468 6344 or 402 592 7077
Fax: 402 592 7969

**Price Index Number:**   **0.82/0.50/1.0**
**Price Index Standard:**   **concrete block/ concrete block + R-19/2x6 wall + R-19**

### POLYSTYRENE CONCRETE FORM SYSTEM

3-10 REWARD WALL is a form system for concrete incorporating lap-jointed polystyrene forms into the finished structure, providing insulation and acoustical absorption. Plastic furring strips are an integral part of the form blocks for screwing on drywall for both basement and above-grade exterior and interior walls. Each form block is 16 in. high by 48 in. long by 9¼ in. or 11 in. thick, resulting in a 6-in. or 8-in. reinforced concrete wall with an R-value of R-22+. Form blocks assemble quickly, are fire resistant, and use 25% to 50% less concrete while creating a 50% stronger, well-insulated wall. No CFCs are used in the manufacture of this product. Approved for use by BOCA and SBCCI.

## 47   Advanced Foam Plastics, Inc.

5250 N. Sherman St., Denver, CO 80216
Phone: 800 525 8697 or 303 297 3844
Fax: 303 292 2613

**Price Index Number:**   **1.40**
**Price Index Standard:**   **poured concrete wall using plywood**

### CONCRETE FORM SYSTEM

DIAMOND SNAPFORM formwork and foundation insulation manufactured from expanded polystyrene with a 10% recycled material content and without CFCs, HCFCs, or formaldehyde. Wall is typically 12 in. thick, including 2 in. of insulation on each side, for an R-value of R-20. Insulation is held in place with polyethylene ties.

## 48 American ConForm Industries, Inc.

1820 S. Santa Fe St., Santa Ana, CA 92705
Phone: 800 266 3676 or 714 662 1100
Fax: 714 662 0405

**Price Index Number:** 1.0
**Price Index Standard:** formed concrete wall
with insulation

### POLYSTYRENE CONCRETE FORM SYSTEM

SMARTBLOCK is a form system for concrete incorporating the expanded polystyrene forms into the finished structure, providing insulation and acoustical absorption. Used for both basement and above-grade exterior and interior walls. Each form block is 10⅔ in. or 12¾ in. high by 40 in. long by 3¾ in.,5¾ in., 7¾ in., or 10 in. deep, resulting in a reinforced concrete wall with an R-value of R-17.6. Forms have a flame spread of 10 while creating an up to 50% stronger, well-insulated wall. No CFCs are used in the manufacture of this product, and it will not deteriorate with age. Uses 13% less concrete than standard formwork.

## 49 American Polysteel Forms

5150F Edith Blvd. N.E., Albuquerque, NM 87107
Phone: 800 977 3676   Fax: 505 345 8154

**Price Index Number:** 0.82/0.50/1.0
**Price Index Standard:** concrete block/
concrete block +
R-19/2x6 wall + R-19

### POLYSTYRENE CONCRETE FORM SYSTEM

POLYSTEEL FORMS CONCRETE WALL SYSTEM is a form system for concrete incorporating tongue-and-groove interlocking polystyrene forms into the finished structure, providing insulation and acoustical absorption. Metal furring strips are an integral part of the form blocks for screwing on drywall for both basement and above-grade exterior and interior walls. Each form block is 16 in. high by 48 in. long by 9¼ in. or 11 in. thick, resulting in a 6-in. or 8-in. reinforced concrete wall with an R-value of R-22+. Form blocks assemble quickly, are fire resistant, and use 25% to 50% less concrete while creating a 50% stronger, well-insulated wall. No CFCs are used in the manufacture of this product. Approved for use by BOCA, ICBO, and SBCCI.

## 50 Composite Technologies

525 E. 2nd St., PO Box 1888, Ames, IA 50010
Phone: 800 232 1748 or 515 232 1748
Fax: 515 232 0800

**Price Index Number:** 1.18
**Price Index Standard:** concrete tilt-up wall

### COMPOSITE FIBER TIES

THERMALMASS is a thermal-wall system for concrete incorporating predrilled Dow Styrofoam panels embedded in concrete walls using composite-fiber wall ties to eliminate any significant thermal bridging. System is designed to take advantage of the thermal mass of the concrete while providing insulation and a moisture seal. Used for both basement and above-grade exterior and interior walls, and tilt-up, cast-in-place, or precast structures. Can be used to form a radius as small as 8 ft. Improvements are being made to the current system to reduce labor by 50%.

## 51 Insteel Construction

2610 Sidney Lanier Dr., Brunswich, GA 31525
Phone: 912 264 3772   Fax: 912 264 3774

**Price Index Number:** 1.30
**Price Index Standard:** framed stud wall

### POLYSTYRENE CONCRETE FORM SYSTEM

INSTEEL is a shotcrete-applied concrete panel system consisting of a space frame-style wire mesh with polystyrene insulation at the center. After concrete is applied to both sides of the wire mesh, it is troweled to desired finish for floors, walls, and ceilings. Works well with hydronic heating systems for accurate placement of piping.

## 52 K and B Associates

PO Box 35605, Monte Sereno, CA 95030
Phone: 800 742 0862 or 408 395 3394
Fax: 800 742 0862 or 408 395 3394

**Price Index Number:** 1.0
**Price Index Standard:** formed concrete wall
with insulation

### POLYSTYRENE CONCRETE FORM SYSTEM

E-Z FORM incorporates interlocking polystyrene forms into the finished structure, providing insulation and acoustical absorption. Used for both basements and above-grade exterior and interior walls. Each form block is 8 in. high by 48 in. long by 10 in. deep, resulting in a 6½-in. reinforced concrete wall with an R-value of R-18. Form blocks assemble quickly, are fire resistant, and use less concrete than standard walls, maximizing the concrete strength by slowing the curing time. Blocks are made for the U.S. market using English units that coordinate well with panel-construction systems.

## 53 Lite-Form, Inc.

1210 Steuban St., Sioux City, IA 51105
Phone: 800 551 3313 or 712 252 3704
Fax: 712 252 3259

**Price Index Number:** 0.95
**Price Index Standard:** poured concrete wall
using plywood

### CONCRETE FORM SYSTEM

LITE-FORM formwork and foundation insulation manufactured from extruded polystyrene. STYRO-FORM is the same as LITE-FORM but preassembled to reduce labor costs. FOLD-FORM is the same as LITE-FORM but made from expanded polystyrene for above-grade stucco applications. Wall is typically 12 in. thick, including 2 in. of insulation on each side, for a minimum R-value of R-20. Insulation is held in place with polypropylene ties having a 20% recycled content. All forms are made from 2 in. thick polystyrene with a density of 2 lbs. for 25 psi.

## 54 Homasote Co.

PO Box 7240, West Trenton, NJ 08628-0240
Phone: 800 257 9491 or 604 883 3300
Fax: 609 530 1584

**Price Index Number:** 0.80
**Price Index Standard:** asphalt impregnated
felt, ½ in.

### EXPANSION-JOINT FILLER

HOMEX 300 is a resilient expansion-joint filler strip that has been premolded and treated for use in concrete and masonry joints. Has been used in airport runways, highways, driveways, and retaining walls. Homasote products are made largely from recycled newsprint. Treated to protect against termites, rot, and fungi. Contains no urea formaldehyde or asbestos.

## 55 M. A. Industries, Inc.

303 Dividend Dr., Peachtree City, GA 30269
Phone: 800 241 8250 or 770 487 7761
Fax: 770 487 1482

**Price Index Number:** 1.0
**Price Index Standard:** cardboard cylinder
mold

### CONCRETE CYLINDER MOLD

Cylinder mold for testing concrete core samples made from recycled plastic from used battery casings.

### 56  W. R. Meadows

PO Box 543, Elgin, IL 60121
Phone: 800 342 5976 or 708 683 4500
Fax: 708 683 4544

**Price Index Number:**  **0.80**
**Price Index Standard:**  **asphalt impregnated felt, ½ in.**

**EXPANSION-JOINT FILLER**
SEALTIGHT self-expanding cork expansion joint filler is formed and compressed under heat and pressure to permit expansion up to 140%. This ability to expand permits the filler to compensate for concrete contraction. Designed to be nonextruding and highly resilient and to effectively maintain contact with joint interfaces. Meets ASTM D1752, Type III; AASHTO M 153, Type III; Corps of Eng. Spec. CRD-C 509, Type III; and FAA Spec. Item P-610-2.7. Cork is a renewable resource harvested from the bark of the cork oak on a 9-year cycle, without destroying the trees.

### 57  W. R. Meadows

PO Box 543, Elgin, IL 60121
Phone: 800 342 5976 or 708 683 4500
Fax: 708 683 4544

**Price Index Number:**  **0.80, curing/ 0.90, sealer**
**Price Index Standard:**  **VOC-based curing compound/ VOC-based sealer**

**CONCRETE ADMIXTURES AND CURING COMPOUNDS**
Manufacturer of water-based GREENLINE ADMIXTURES for concrete, curing compounds, accelerators, bonding agents, and sealers, which meet California and EPA regulations for VOCs.

### 58  Concrete Designs, Inc.

3650 S. Broadmont, Tucson, AZ 85713
Phone: 800 279 2278 or 602 624 6653
Fax: 602 624 3420

**Price Index Number:**  **0.10**
**Price Index Standard:**  **carved stone**

**ORNAMENTAL PRECAST CONCRETE**
Ornamental precast concrete moldings, mantels, balustrades, chimney caps, and columns. Largest selection is in the true classical beaux arts and Spanish colonial styles. Architects will appreciate the presentation of classical forms in the product catalog. Custom orders accepted. A limited selection of columns, site furnishings, and window and door surrounds are also available in FIBERCAST, a blend of alkali-resistant glass fibers, cement, and mortar. FIBERCAST is 65% to 80% lighter weight than concrete, with a higher flexural strength and a smoother texture.

### 59  Garland-White and Co.

1 Tara Ct., Union City, CA 94587
Phone: 510 471 5666   Fax: 510 471 3583

**Price Index Number:**  **1.3**
**Price Index Standard:**  **standard thin-set mortar cement**

**THIN-SET MORTAR**
BON-DON is a thin-set white mortar cement tile setting material, without sand or latex, which has been used for 30 years. Manufacturer claims that it works well for people with a chemical hypersensitivity to standard thin-set mortar cement.

## 60  Cunningham Brick Co., Inc.

Rt. 2, Cunningham Brick Rd., Thomasville, NC 27360
Phone: 910 472 6181
Fax: 910 224 0002 or 910 472 6181

**Price Index Number:**  1.0
**Price Index Standard:**  brick

### BRICK

Bricks manufactured from reprocessed oil-contaminated soil. Available as solid or face bricks in 3 sizes: modular, 7⅝ in. by 3⅝ in. by 2¼ in. (face); engineer, 7⅝ in. by 3⅝ in. by 2¾ in. (face); and queen, 7⅝ in. by 2¾ in. by 2¾ in. Color is red. Distribution is typically limited to Eastern Canada down to Florida and west to Texas.

## 61  Durox Building Units Ltd.

Northumberland Rd., Linford, Stanford-le-Hope
Essex SS17 OPY England
Phone: 44 1375 673344   Fax: 44 1375 360647

**Price Index Number:**  2.0 material/
1.3 installed
**Price Index Standard:**  concrete block

### AERATED CONCRETE BLOCK

Durox autoclaved aerated concrete blocks and reinforced headers are designed for use in both load-bearing and non-load-bearing walls. Manufactured from cement, sand, lime and water with aluminum powder added in small amounts. A controlled chemical reaction, which generates hydrogen bubbles, produces a durable, lightweight cellular structure that can be worked with ordinary tools. These products are steam-cured in autoclaves. Weight is 31 lbs./cu. ft., and it has a noncombustible, Class O flame spread rating and a fire resistance of 1-hour per 1 in. thickness of wall. A 4-in.-thick wall has an R-value of 5. Company recycles waste materials in its manufacturing process.

## 62  Hebel U.S.A. LP

2408 Mt. Vernon Rd., Atlanta, GA 30338
Phone: 800 354 3235 or 770 394 5546
Fax: 770 394 5564

**Price Index Number:**  1.02-1.03
**Price Index Standard:**  concrete block

### AERATED CONCRETE BLOCK

Autoclaved aerated concrete products for wall slabs, floor slabs, flat or sloping roof, and ceiling slabs. Manufactured from quartz, cement, sand, lime, and water with an aluminum paste added. A controlled chemical reaction, which generates hydrogen bubbles, produces a durable lightweight cellular structure that can be worked with ordinary tools. These products are steam-cured in autoclaves. Blocks measure 20 in. long by 10 in. wide by 2 in. to 16 in. thick. An 8-in.-thick block has an R-value of 10. Custom sizes can be ordered. Floor, roof, and wall slabs come as long as 20 ft. by 2 ft. wide by 4 in. to 12 in. thick. Weight is about one-fifth of standard concrete and uses less cement. Fire rating is noncombustible. Blocks are not sensitive to frost and will not expand or shrink. The manufacturing process and the final product does not produce or contain any pollutants or toxic substances. Distribution limited from Washington D.C., to Texas, including Forida and Georgia.

## 63  ENER-GRID, Inc.

6847 S. Rainbow Rd., Buckeye, AZ 85326
Phone: 800 715 5486 or 602 386 2232
Fax: 602 386 3298

**Price Index Number:**  **1.03-1.05**
**Price Index Standard:**  **conventional residential construction**

### POLYSTYRENE/CEMENT FORMS

The ENER-GRID building system element consists of 87% recycled post-consumer polystyrene beads with 13% cementitious binder. The polystyrene is 100% recycled. Lightweight (180-lb./10-ft.-long element), versatile building element that can be formed in many shapes. ENER-GRID elements are stacked and filled with concrete to create a structural concrete grid. Stucco directly applied to surface. Average element is 8 in.,10 in.,12 in. thick by 15 in. high by 10 ft. with an R-value of R-36.44. Being tested for ICBO approval. Structural engineering calculations are available. Fire rating is 2 hours.

## 64  Insulated Masonry Systems, Inc.

7234 E. Shoeman Ln., Suite 1, Scottsdale, AZ 85251
Phone: 602 970 0711   Fax: 602 970 1243

**Price Index Number:**  **1.50/0.70**
**Price Index Standard:**  **2x4/2x6 framed stud wall**

### POLYSTYRENE/CEMENT BLOCK

The IMSI WALL BLOCK SYSTEM is a combination of concrete block and expanded polystyrene (EPS) inserts and surface bonding cements on both sides to produce an insulated masonry wall. For stem and foundation walls, the blocks come in two sizes: 8 in. by 8 in. by 16 in. and 12 in. by 8 in. by 16 in. R-value is R-19 for 8 in. thickness and R-30 for 12 in. Fire rating is 2-4 hours. Block is impervious to termites and vermin and offers good acoustical isolation. Manufacturer claims it performs well in earthquake zones.

## 65  InteGrid Building Systems

PO Box 5298, Berkeley, CA 94705-0298
Phone: 510 845 1100   Fax: 510 845 6886

**Price Index Number:**  **0.80/0.96**
**Price Index Standard:**  **concrete block/ stick framing**

### POLYSTYRENE/CEMENT BLOCK

RASTRA block is a forming system that creates a reinforced concrete post and lintel structural frame. The basic block consists of a blend of recycled polystyrene beads coated with cement, forming hollow cores. The cores are filled with steel-reinforced concrete, creating an overall insulated wall thickness of 8½ in., 10 in., 12 in., and 14 in. with an R-24 insulation value. 5-hour fire rating. Also, excellent acoustical qualities reducing sound by 53 dB.

## 66  Midwest Faswall, Inc.

404 N. Forrest Ave., Ottumwa, IA 52501
Phone: 888 682 1212 or 515 682 1212
Fax: 515 683 1212

**Price Index Number:**  **1.15**
**Price Index Standard:**  **framed stud wall**

### WOOD/CONCRETE FORMS

FASWALL Concrete Systems are wood/concrete composite forms, stacked dry, and filled with reinforced concrete for walls and foundations. FASWALL creates a "breathable" wall that does not require building paper. Stucco is directly applied to the wall surface, and the interior surface can be finished with a lime slurry. Weight of the forms is 36 lbs./cu. ft. Forms can be cut, routed, and drilled with ordinary tools. 50-year use in Europe. 10-year use in Canada. Claims: R-11 up to R-24 with 3-in. insulation; sound absorptive; vermin, frost, and rot resistant. Forms are 12 in. by 12 in., 24 in. or 36 in. long by 9 in. wide. A 9-in.-thick wall has an R-value of R-12 and a fire rating of 4 hours. Manufacturer claims this product to be nontoxic.

## 67 Sparfil Blok Florida, Inc.

PO Box 270336, Tampa, FL 33688
Phone: 813 963 3794   Fax: 813 963 3794

**Price Index Number:** 0.1-0.13, installed
**Price Index Standard:** concrete block

### POLYSTYRENE/CEMENT BLOCK

WALL SYSTEM II consists of insulated concrete blocks (expanded polystyrene beads in concrete mixture with fly ash), used with surface-bonding material on both sides. Blocks are dry stacked with a surface-bonding finish material in a fibrous mesh that creates a waterproof, airtight wall structure. 40% lighter than standard block. Block widths from 4 in. to 12 in. in 2-in. increments. 12-in. block R-value is R-12. With optional polystyrene insert, R-value is R-24.5. Can use reground polystyrene, but limited by available sources. Variety of finishes available.

## 68 Superlite Block

4150 W. Turney, Phoenix, AZ 85019
Phone: 800 366 7877 or 602 352 3500
Fax: 602 352 3814

**Price Index Number:** 0.52
**Price Index Standard:** concrete block, 8 in. by 8 in. by 16 in.

### CONCRETE BLOCK

INTEGRA block wall system is made from masonry block units with a reduced center web. The reduced web allows polyurethane foam to be blown in after the wall is constructed. Walls are post-tensioned with vertical reinforcing. Available in 6-in. by 8-in. by 16-in. and 8-in. by 8-in. by 16-in. sizes. R-value for 6-in. wall is R-23 and 8 in. wall is R-28. Fire rating is 1 hour for 6 in. blocks and 2 hours for 8-in. blocks.

## 69 Conklin's Authentic Antique Barnwood

R.R. 1, Box 70, Susquehanna, PA 18847
Phone: 717 465 3832   Fax: 717 465 3832

**Price Index Number:** 1.0
**Price Index Standard:** quarried bluestone

### FIELD STONE

Reclaimed Pennsylvania wall stone for interior and exterior walls and floors. 3 in. to 6 in. thick. Color is blue-gray. Reclaimed from old walls erected by the early settlers.

## 70 Quinstone Industries, Inc.

PO Box 1026, 1112 W. King St., Quincy, FL 32351
Phone: 800 621 0565 or 904 627 1083
Fax: 904 627 2640

**Price Index Number:** 0.50/0.10
**Price Index Standard:** carved stone + installation labor/ stone fireplace mantels

### MOLDED CELLULOSE SURFACES

Molded cellulose surfaces produced from nontoxic, recycled cellulose to simulate both rough and carved stone. Products include flat panels, fireplace surrounds, and decorative pieces, such as various heads, animals, and plinth blocks. Available in 7 colors and 4 stone textures: travertine, dolomite, Texas limestone, and coral. Can be worked with conventional hand tools, glue, and nails. Manufactured from 15% recycled waste paper, gypsum, resin and nontoxic color. Nonflammable, flame-spread rating of 0. Currently for interior use, but the company is developing a high-grade cellulose and resin product for exterior use.

## 71  Syndesis Studio

2908 Colorado Ave., Santa Monica, CA 90404-3616
Phone: 310 829 9932   Fax: 310 829 5641

**Price Index Number:**   **1.0**
**Price Index Standard:**   **granite**

### PRECAST CONCRETE SURFACES
SYNDECRETE is a lightweight, precast concrete surfacing material used in tabletops, countertops, sinks, tub surrounds, tile flooring, site works, and furniture. Aggregates can be added to vary the texture and appearance. Made to custom order. Can utilize recycled materials, including stone, wood, steel shavings, and plastic chips for unique patterns and designs. At 75 to 85 lbs./cu. ft., the weight is one-half that of standard concrete. SYNDECRETE can be worked with standard woodworking tools. Fly ash, volcanic ash, and polypropylene fibers in the mix make it more resistant to cracking and chipping than standard concrete. For countertop applications, sink cutouts can be made on-site. Appropriate for interior or exterior use.

## METAL MATERIALS                                                 05010

## 72  C F and I Steel LP

PO Box 316, Pueblo, CO 81002
Phone: 800 992 9900 or 719 561 6000
Fax: 719 561 6862 or 719 561 6180

**Price Index Number:**   **1.0**
**Price Index Standard:**   **steel from iron ore**

### WIRE AND TUBULAR STEEL
Manufacturer of steel products from 100% recycled scrap metal. Products include nails, rolled wire, tubular steel, and rails. A subsidiary of Oregon Steel Mills.

## STRUCTURAL METAL FRAMING                                        05100

## 73  Angeles Metal Systems

4817 E. Sheila St., Los Angeles, CA 90040
Phone: 800 366 6464 or 213 268 1777
Fax: 213 268 8996

**Price Index Number:**   **0.70-1.30, steel studs**
**Price Index Standard:**   **wood studs**

### STEEL STUDS
Fabricator of steel framing studs, joists, and trusses for residential and commercial construction. Claims 60% to 70% pre- and post-consumer recycled steel.

## 74  Tri-Steel Structures

5400 S. Stemmons (I-35E), Denton, TX 76205
Phone: 800 874 7833 or 817 497 7070
Fax: 800 874 7335 or 817 497 7497

**Price Index Number:**   **0.95**
**Price Index Standard:**   **residential stick framing**

### STEEL STUDS
Complete metal framing systems for both residential and commercial structures. Framing is based on 8 ft. o.c. modules. Product is manufactured from about 70% recycled steel from cars.

### 75  Vulcraft Steel Joist and Girders

PO Box 637, Brigham City, UT 84302
Phone: 801 734 9433    Fax: 801 723 5423

**Price Index Number:**    **1.0**
**Price Index Standard:**   **steel from iron ore**

**STEEL JOISTS AND GIRDERS**
Fabricator of steel joists and girders using 100% recycled steel content supplied by parent company and manufacturer Nucor Corporation. Girders and joists are custom engineered for projects.

### 76  Advanced Framing Systems, Inc.

PO Box 440, 882 Main St. #130, Conyers, GA 30207
Phone: 800 633 8600 or 770 483 8100
Fax: 770 483 0014

**Price Index Number:**    **1.05-1.06**
**Price Index Standard:**   **residential stick framing**

**STEEL FRAMING**
Complete metal framing systems for both residential and commercial structures. Exterior wall systems are designed with thermal break to avoid heat loss through steel studs, resulting in an 8-in.-thick finished wall. Framing is based on 4 ft. o.c.

### 77  Vanport Steel and Supply

1403 N.E. 106th St., Vancouver, WA 98686
Phone: 800 288 9279 or 206 573 9010
Fax: 206 573 8842

**Price Index Number:**    **1.05-1.10/1.02-1.05**
**Price Index Standard:**   **wood framing/ overall project**

**STEEL FRAMING**
PIONEER STEEL FRAMING is a steel framing package on a 2-ft. module for residential and small commercial projects. Structural members are custom engineered for each project. Recycled content of steel products is 66%.

### 78  Neenah Foundry Co.

PO Box 729, Neenah, WI 54957
Phone: 800 558 5075 or 414 725 7000
Fax: 414 729 3682

**Price Index Number:**    **1.0**
**Price Index Standard:**   **metal products from ore**

**METAL CASTINGS**
Manufacturer of cast-metal products, which include tree grates and guards, trench drains, and manhole covers. Cast iron products have a 99% recycled content. Aluminum and bronze products are special-order items.

## 79  Aged Woods, Inc.

2331 E. Market St., York, PA 17402
Phone: 800 233 9307 or 717 840 0330
Fax: 717 840 1468

**Price Index Number:**  **2.20**
**Price Index Standard:**  **red oak flooring**

**RECYCLED LUMBER**
Recycled lumber salvaged from barns. Kiln-dried barnwood paneling, stair parts, moldings, cabinets, and beams. Also available is ¾-in. tongue-and-groove wood flooring.

## 80  Allied Demolition, Inc.

7901 US Highway 85, Commerce City, CO 80022
Phone: 303 289 3366   Fax: 303 289 3543

**Price Index Number:**  **0.5**
**Price Index Standard:**  **lumber**

**RECYCLED LUMBER**
Sale of distressed wood salvaged from lumberyards. Species available include hemlock fir, Douglas fir, cedar, pine, and redwood. Must be picked up at the yard.

## 81  Bear Creek Lumber Co.

PO Box 669, Winthrop, WA 98862
Phone: 509 997 3110   Fax: 509 997 2040

**Price Index Number:**  **1.0**
**Price Index Standard:**  **lumber from**
**non-sustainable**
**sources**

**SUSTAINABLE HARVEST**
Supplies lumber from sustainably harvested sources.

## 82  Big Creek Lumber Co.

3564 Highway 1, Davenport, CA 95017
Phone: 800 464 2770   Fax: 408 423 2800

**Price Index Number:**  **1.0**
**Price Index Standard:**  **lumber from**
**non-sustainable**
**sources**

**SUSTAINABLE HARVEST**
Selectively cut from sustainably managed sources in the Santa Cruz mountains. Wood is typically redwood with some Douglas fir.

## 83  Bronx 2000/Big City Forest

1809 Carter Ave., Bronx, NY 10457
Phone: 718 731 3931   Fax: 718 583 2047

**Price Index Number:**  **1.0**
**Price Index Standard:**  **butcher-block**
**furniture**

**RECYCLED LUMBER**
Manufacturer collects discarded wood pallets, crates, skids, chocks, and packaging materials from New York City businesses and remills the lumber into pallets, butcher-block furniture, and flooring. Products available in oak, cherry, maple, poplar, mahogany, pine, and fir.

## 84  Caldwell Building Wrecking

195 Bayshore Blvd., San Francisco, CA 94124
Phone: 415 550 6777    Fax: 415 550 0349

**Price Index Number:    1.0**
**Price Index Standard:  Douglas fir lumber**

**RECYCLED LUMBER**
Recycled lumber salvaged from demolition of both residential and commercial buildings.

## 85  Centre Mills Antique Wood

PO Box 16, Aspers, PA 17304
Phone: 717 334 0249    Fax: 717 334 6223

**Price Index Number:    2.6, 5-in. to 12-in.-
                         wide antique
                         oak planks**
**Price Index Standard:  red oak flooring**

**RECYCLED LUMBER**
Recycled antique wood salvaged from old buildings. Wood includes hand-hewn beams, logs from houses, floor planks, and wallboards plus a variety of other house and barn components. Documentation provided with wood.

## 86  Collins Pine Co.

1618 S.W. 1st Ave., Suite 300, Portland, OR 97201
Phone: 800 329 1219 or 503 227 1219
Fax: 503 227 5349

**Price Index Number:    1.05**
**Price Index Standard:  lumber from
                         non-sustainable
                         sources**

**SUSTAINABLE HARVEST**
Supplies hardwoods and softwoods from sustainably harvested sources. Also produces veneers of cherry, red oak, white oak, ash, and maple (hard and soft). Takes custom orders. Runs sawmills in California, Oregon, and Pennsylvania and ships anywhere.

## 87  Conklin's Authentic Antique Barnwood

R.R. 1, Box 70, Susquehanna, PA 18847
Phone: 717 465 3832    Fax: 717 465 3832

**Price Index Number:    8.2, 4x4/0.94, 6x12**
**Price Index Standard:  Douglas fir lumber**

**RECLAIMED LUMBER**
Pine, hemlock, chestnut, oak, and heart pine hand-hewn beams and barn-wood siding from 19th-century barns.

## 88  Crossroads Recycled Lumber

PO Box 184, O'Neals, CA 93645
Phone: 209 868 3646    Fax: 209 868 3646

**Price Index Number:    0.40-1.0**
**Price Index Standard:  Douglas fir #1 or
                         better lumber**

**RECYCLED LUMBER**
Recycled lumber salvaged from industrial sources such as buildings, bridges, and railroad lines. Product includes raw and remilled lumber of Douglas fir and redwood species.

## 89 Cut and Dried Hardwoods

241 S. Cedros Ave., Solana Beach, CA 92075
Phone: 619 491 0442   Fax: 619 481 1703

**Price Index Number:** **1.20**
**Price Index Standard:** **non-sustainably**
**harvested hardwood**

**SUSTAINABLE HARVEST**
Supplies hardwoods from sustainably managed sources in Plan Piloto, Mexico. Endorsed by the Rainforest Action Network.

## 90 Duluth Timber Co.

PO Box 16717, 3310 Minnesota Ave.,
Duluth, MN 55816-0717
Phone: 218 727 2145   Fax: 218 727 0393

**Price Index Number:** **1.20**
**Price Index Standard:** **Douglas fir lumber**

**RECYCLED LUMBER**
Recycled wood salvaged from buildings. Wood includes beams, trim, flooring, and stair components. Available in Douglas fir, redwood, and southern yellow pine.

## 91 Edensaw Woods Ltd.

211 Seton Rd., Port Townsend, WA 98368
Phone: 800 950 3336 or 360 385 7878
Fax: 360 385 5215

**Price Index Number:** **1.0**
**Price Index Standard:** **non-sustainably**
**harvested hardwood**

**SUSTAINABLE HARVEST**
Distributors for a variety of tropical hardwoods from the Yanesha Co-op, a small Peruvian cooperative using natural forest management practices, and Plan Piloto, a Mexican sustained yield program. Also supplies domestic hardwoods such as cherry, red and white oak, maple, and ash. Plywood from sustainably managed sources in Africa is also available. Endorsed by the Rainforest Action Network.

## 92 G. R. Plume Co.

1301 Meadow Ave., Suite B-11 and 12,
Bellingham, WA 98226
Phone: 360 676 5658   Fax: 360 734 1909

**Price Index Number:** **1.5**
**Price Index Standard:** **Douglas fir lumber**

**RECYCLED LUMBER**
Recycled timbers, flooring, paneling, and architectural millwork salvaged from century-old structures. Wood is Douglas fir. This is a full-service mill shop that works from custom orders. This is the only shop in the country certified by the APA to glue up heavy timbers without requiring lag bolts for long spans.

## 93 Gilmer Wood Co.

2211 N.W. Saint Helens Rd., Portland, OR 97210
Phone: 503 274 1271   Fax: 503 274 9839

**Price Index Number:**   **1.0**
**Price Index Standard:**   **non-sustainably harvested hardwood**

### SUSTAINABLE HARVEST
Supplies hardwoods from sustainably managed sources in Plan Piloto, Mexico, and lacewood from plantations in Brazil. Also, imports mill waste shorts in zebrawood and other hardwoods at a 25% discount.

## 94 Goodwin Heart Pine Co.

106 S.W. 109th Pl., Micanopy, FL 32667
Phone: 800 336 3118 or 352 466 0339
Fax: 352 466 0608

**Price Index Number:**   **2.0-2.6, reclaimed heart pine flooring**
**Price Index Standard:**   **red oak floor**

### RECYCLED LUMBER
Beams, flooring, and molding milled from virgin cypress and heart pine logs reclaimed from logs that sank while being floated downstream to mills in the late 19th century. Cold water and lack of oxygen has preserved the logs for a century. Also, the company remills lumber salvaged from old-buildings, mostly old growth heart pine.

## 95 Handloggers Hardwood

135 E. Francis Drake Blvd., Larkspur, CA 94939
Phone: 800 461 1969 or 415 461 1180
Fax: 415 461 1187

**Price Index Number:**   **1.0**
**Price Index Standard:**   **non-sustainably harvested hardwood**

### SUSTAINABLE HARVEST
Distributors for a variety of tropical hardwoods from the Yanesha Co-op, a small Peruvian cooperative using natural forest-management practices.

## 96 Institute for Sustainable Forestry

PO Box 1580, Redway, CA 95560
Phone: 707 923 4719   Fax: 707 923 4257

**Price Index Number:**   **1.0**
**Price Index Standard:**   **non-sustainably harvested hardwood**

### SUSTAINABLE HARVEST
Puts potential customers in contact with suppliers and buyers of native domestic hardwoods and softwoods. Nonprofit corporation. Member of Smart Wood Network.

## 97  Into the Woods

300 N. Water St., Petaluma, CA 94952
Phone: 707 763 0159
Fax: 707 763 0159 or 707 769 8952

**Price Index Number:   n/a**
**Price Index Standard:  n/a**

**RECYCLED LUMBER**
Various exotic lumber from urban sources such as forests and
orchards in addition to native and local hardwoods. Recycled
lumber is also available. Typical uses include countertops,
cabinets, baseboards, casings, doors, windows, flooring, and
molding. Available in a variety of wood species including
fruit woods, and substitutes for rare woods such as black
locust for teak or acacia for rosewood.

## 98  J Squared Timberworks, Inc.

449 N. 34th St., Seattle, WA 98103
Phone: 800 598 3074 or 206 633 0504
Fax: 206 633 0565

**Price Index Number:   1.5**
**Price Index Standard:  Douglas fir lumber**

**RECYCLED LUMBER**
Recycled and remilled lumber salvaged from old lumber mill
buildings, barns, and warehouses. Wood is old-growth
Douglas fir in various sizes. J Squared also provides a design
and construction service for timber-framed houses.

## 99  Jefferson Recycled Woodworks

PO Box 696, 1104 Firenze St., McCloud, CA 96057
Phone: 916 964 2740   Fax: 916 964 2745

**Price Index Number:   0.7**
**Price Index Standard:  virgin wood of**
**                       similar quality**

**RECYCLED LUMBER**
Recycled and remilled lumber salvaged from old structures
located in the Pacific Northwest. Orders are custom milled to
specifications for such uses as flooring, framing, or trim.
Available as beams, timbers, flooring, paneling, and finish
lumber of redwood, pine, Douglas fir, Port Orford cedar, and
other woods.

## 100  Larson Wood Products, Inc.

31421 Coburg Bottom Loop, Eugene, OR 97401
Phone: 541 988 9155   Fax: 541 343 3279

**Price Index Number:   1.0**
**Price Index Standard:  veneer from non-**
**                       sustainable source**

**SUSTAINABLE HARVEST**
The company owns and operates an old Georgia-Pacific mill
in Brazil that has been converted to a sustainable-harvest
operation. Supplies plywood substrate and veneer cores
mostly to Georgia-Pacific and Plywood Tropics USA for
manufacturing finished product. Will also accept orders from
other manufacturers. Working with the Rainforest Alliance.

## 101  Maxwell Pacific

PO Box 4127, Malibu, CA 90264
Phone: 310 457 4533   Fax: 310 457 8308

**Price Index Number:   1.0**
**Price Index Standard:  virgin wood of**
**                       similar quality**

**RECYCLED LUMBER**
Recycled salvaged lumber custom milled for flooring,
paneling, molding, siding, and decking. Wood available in
a variety of species including Douglas fir, redwood, cedar,
and pine.

## 102 Menominee Tribal Enterprises

PO Box 10, Neopit, WI 54150
Phone: 715 756 2311 or 715 799 3896 (mill)
Fax: 715 756 2386

**Price Index Number:** 1.0
**Price Index Standard:** non-sustainably
harvested hardwood

### SUSTAINABLE HARVEST
The Menominee are a tribe of American Indians who have been practicing sustained-yield management on 220,000 acres of their forest land in northcentral Wisconsin since 1854. Eastern hardwoods available include hard maple, ash, birch, and white pine. Smart Wood and Scientific Certification Systems certified.

## 103 Mount Storm

7890 Bell Rd., Windsor, CA 95492
Phone: 707 838 3177    Fax: 707 838 4413

**Price Index Number:** 1.0
**Price Index Standard:** non-sustainably
harvested hardwood

### SUSTAINABLE HARVEST
Distributors for a variety of tropical hardwoods from the Yanesha Co-op, a small Peruvian cooperative using natural forest-management practices.

## 104 Mountain Lumber Co.

PO Box 289, Ruckersville, VA 22968
Phone: 800 445 2671 or 804 985 3646
Fax: 804 985 4105

**Price Index Number:** 2.0
**Price Index Standard:** red oak flooring

### RECYCLED LUMBER
Recycled and remilled long-leaf heart pine salvaged from late 19th-century structures for flooring, stair parts, beams, paneling, wainscots, and moldings. Heart pine comes in 7 grades stocked in ¾-in. or 1-in. nominal thicknesses and 3-in. to 10-in. widths. Other woods available include American oak, chestnut, and French oak salvaged from old French boxcars. Will custom mill to order.

## 105 Natural Resources

PO Box 157, Petrolia, CA 95558
Phone: 707 629 3679    Fax: 707 629 3679

**Price Index Number:** 1.0
**Price Index Standard:** redwood

### RECYCLED LUMBER
Remilled lumber salvaged from demolition of buildings. Available wood species include redwood, Douglas fir and Alaskan yellow cedar. Certified ecologically harvested lumber also available.

## 106 Peter Lang Co.

3115 Porter Creek Rd., Santa Rosa, CA 95404
Phone: 800 616 2695 or 707 579 1341
Fax: 707 579 8777

**Price Index Number:** 1.0
**Price Index Standard:** **non-sustainably harvested hardwood**

### SUSTAINABLE HARVEST
Small mill with connections to various logging companies to be able to remove selected hardwoods prior to main logging operation. These are trees that would typically be used for chips by the larger companies. Available species are black walnut, big-leaf maple, bay laurel, redwood, and madrone. Also available are the burls from these species.

## 107 Pioneer Millworks

1755 Pioneer Rd., Shortsville, NY 14548
Phone: 800 951 9003 or 716 289 3090
Fax: 716 289 3221

**Price Index Number:** 1.5/0.90
**Price Index Standard:** **Douglas fir/long beams**

### RECYCLED LUMBER
Recycled and remilled lumber salvaged from commercial and industrial structures, river bottoms, and wine vats. Beams, flooring, casings, and moldings are available in a variety of sizes and species. Custom milling orders are accepted. Will provide a quote from a faxed material list. Support and consultation service also available for design and construction crew for timber-frame buildings.

## 108 Pittsford Lumber Co.

50 State St., Pittsford, NY 14534
Phone: 716 381 3489   Fax: 716 586 1934

**Price Index Number:** 1.0
**Price Index Standard:** **non-sustainably harvested hardwood**

### SUSTAINABLE HARVEST
Distributors for a variety of tropical hardwoods from Mexico and Central America that have been certified as "Smart Wood" by Rainforest Alliance. These include the Yanesha Co-op, Mocimboa Da Praia, and Gorongosa Biosphere Reserves in Mozambique.

## 109 Resource Woodworks, Inc.

627 E. 60th, Tacoma, WA 98404
Phone: 206 474 3757   Fax: 206 474 1139

**Price Index Number:** 0.45
**Price Index Standard:** **Douglas fir #1 or better lumber**

### RECYCLED LUMBER
Recycled lumber salvaged from industrial sources such as buildings, bridges, and railroads. Typically old-growth Douglas fir. Custom millwork provided.

## 110 Sierra Timber Framers

PO Box 595, Nevada City, CA 95959-0595
Phone: 916 292 9449

**Price Index Number:** 1.10, timber-framed house
**Price Index Standard:** **conventional-framed custom house**

### RECYCLED LUMBER
Has a stockpile of reclaimed lumber from mills in the Pacific Northwest and access to other sources. Provides residential timber-framing service. Claims less wood is used than in conventional framing. Able to combine nonstructural panelized construction with timber framing for greater thermal efficiency than conventional framing. Also, has expertise in building with bamboo.

## 111  Stein and Collett, Inc.

PO Box 4065, McCall, ID 83638
Phone: 208 634 5374   Fax: 208 634 8228

**Price Index Number:**   **1.0**
**Price Index Standard:**  **Douglas fir lumber**

**RECYCLED LUMBER**
Recycled lumber salvaged from building demolition. Products include beams of Douglas fir, paneling, doors, architectural millwork, and stair systems.

## 112  Tosten Brothers Lumber

PO Box 156, Miranda, CA 95553
Phone: 707 943 3093   Fax: 707 943 3665

**Price Index Number:**   **1.10**
**Price Index Standard:**  **lumber from non-sustainable sources**

**SUSTAINABLE HARVEST**
Selectively cut from sustainably managed sources for custom orders. Wood is typically high-quality Douglas fir or redwood.

## 113  Tree Products Hardwoods

PO Box 772, Eugene, OR 97440
Phone: 541 689 8515   Fax: 541 688 4924

**Price Index Number:**   **1.5**
**Price Index Standard:**  **non-sustainably harvested hardwood**

**SUSTAINABLE HARVEST**
Exotic woods from sustainably harvested sources in South America.

## 114  Under the Canopy Wood Products

5500 Nicasio Valley Rd., Nicasio, CA 94946
Phone: 415 662 2472

**Price Index Number:**   **0.8**
**Price Index Standard:**  **virgin clear-heart redwood**

**SUSTAINABLE HARVEST**
Searches for and mills only dead redwood trees felled by nature. Minimum orders of 400 bd. ft. to 1000 bd. ft. Family-run business. Wood is typically of a high quality.

## 115  Wesco Used Lumber

PO Box 1136, El Cerrito, CA 94530
Phone: 510 235 9995

**Price Index Number:**   **1.0**
**Price Index Standard:**  **virgin lumber**

**RECYCLED LUMBER**
Recycled wood salvaged from old buildings. Mostly Douglas fir beams and joists, plus hardwood flooring. Purchase and sell salvaged lumber.

## 116  Wild Iris Forestry

PO Box 1423, Redway, CA 95560
Phone: 707 923 2344   Fax: 707 923 4257

**Price Index Number:**   **1.0**
**Price Index Standard:**   **non-sustainably
harvested lumber**

### RECYCLED LUMBER
Domestic lumber from sustainably managed forests. Works in collaboration with The Institute for Sustainable Forestry, also of Redway, Calif.

## 117  Wildwoods Co.

1055 Samoa Blvd., Arcata, CA 95521
Phone: 707 822 9541   Fax: 707 822 8359

**Price Index Number:**   **1.0**
**Price Index Standard:**   **non-sustainably
harvested hardwood**

### SUSTAINABLE HARVEST
Supplies hardwoods from sustainably managed sources in Plan Piloto, Mexico. Prefers to deal in container lots (7,000 bd. ft. to 7,500 bd. ft.) of green wood. Endorsed by the Rainforest Action Network.

## 118  Wood Floors, Inc.

PO Box 1522, Orangeburg, SC 29116-1864
Phone: 803 534 8478   Fax: 803 533 0051

**Price Index Number:**   **1.3-2.0**
**Price Index Standard:   new heart pine**

### RECYCLED LUMBER
Recycled lumber salvaged from 19th-century buildings for wall paneling, molding, special architectural millwork, and beams from heart pine lumber. Available millwork includes table tops, bar stools, molding, and stair treads and risers.

## 119  Wooden Workbench

202b Airpark Dr., Fort Collins, CO 80524
Phone: 970 484 2423   Fax: 970 482 1572

**Price Index Number:**   **1.0**
**Price Index Standard:   non-sustainably
harvested hardwood**

### SUSTAINABLE HARVEST
Tropical woods from sustainable sources, including teak and maribara from Samoa and Ifelele from Bolivia.

## 120  Woodhouse

PO Box 7336, Rocky Mount, NC 27804
Phone: 919 977 7336   Fax: 919 641 4477

**Price Index Number:**   **1.35, flooring/
5.6, beams**
**Price Index Standard:   red oak flooring/
Douglas fir beams**

### RECYCLED LUMBER
Architectural millwork manufactured from salvaged antique woods. Products include stair parts, moldings, cabinets, and long-leaf heart pine beams and flooring. Flooring remilled from timbers salvaged from abandoned factories and textile mills. Wood is southern yellow pine in standard sizes of ¾-in. thickness, and widths of 3 in. to 9 in. Wood is available in plainsawn or quartersawn styles, in several grades.

## 121  Woodworker's Source

5402 S. 40th St., Phoenix, AZ 85040
Phone: 800 423 2450 or 602 437 4415
Fax: 602 437 3819

**Price Index Number:**   1.0
**Price Index Standard:**   non-sustainably
                            harvested wood

### SUSTAINABLE HARVEST
Distributor for a variety of tropical hardwoods from sources using natural forest-management practices. Woods include teak and maribara from Samoa and ifelele from Bolivia. Also supply FLEXWOOD veneers in many different domestic and exotic species as an alternative to using expensive solid woods. Sheets are laminated to a nonwoven backer and processed to make them flexible.

## 122  Franklin International

2020 Bruck St., Columbus, OH 43207
Phone: 800 877 4583 or 614 443 0241
Fax: 614 445 1555

**Price Index Number:**   1.1
**Price Index Standard:**   panel adhesive

### CONSTRUCTION ADHESIVES
TITEBOND SOLVENT-FREE CONSTRUCTION ADHESIVE has 5.1 grams per liter of VOCs. Applicable to subflooring and meets standard adhesive performance. Manufacturer claims this product to be nontoxic.

## 123  Maze Nails

Dept. JLC, PO Box 449, Peru, IL 61345
Phone: 800 435 5949    Fax: 815 223 7585

**Price Index Number:**   1.30
**Price Index Standard:**   common steel nail

### NAILS
MAZE NAILS made from scrap steel. Recycling in manufacturing process includes reclaiming acid used for cleaning scrap, sorting and selling waste wire, and reclaiming and recycling waste zinc by-products of galvanizing. MAZE NAILS are double-hot-dipped galvanized for greater durability and life than common steel nails.

## 124  Sure Fit Shims

PO Box 35225, Greensboro, NC 27425-5225
Phone: 800 225 0848    Fax: 919 668 7790

**Price Index Number:**   1.10-1.20
**Price Index Standard:**   cedar shims

### PLASTIC SHIMS
Shims manufactured from recycled plastic for construction applications. Available in 9 in. and 3½ in. lengths by 1½ in. width and up to ½ in. maximum thickness. The 9-in. and 3½-in. shims are used in tandem to provide up to a ¾-in. adjustment.

## 125 Compak America

3612 S. 5100 W., Hooper, UT 84315
Phone: 801 965 0979  or 801 731 8451
Fax: 801 731 8517

**Price Index Number:** **1.0**
**Price Index Standard:** **particleboard**

**PARTICLEBOARD**
COMPAKBOARD is manufactered by steam heating and pressing together with a polyurethane resin binder a composite panel using a straw agricultural by-product as a replacement for particleboard. Comes in ¼ in. to ⅝ in. thicknesses. There's no formaldehyde in the product. 17% lighter than conventional particleboard, according to manufacturer.

## 126 Homasote Co.

PO Box 7240, West Trenton, NJ 08628-0240
Phone: 800 257 9491 or 609 883 3300
Fax: 609 883-3300

**Price Index Number:** **1.9**
**Price Index Standard:** **particleboard**

**LAMINATED SHEATHING**
440 CARPETBOARD and COMFORTBASE are compressed paper boards from 100% recycled newsprint cellulose. Both are available as 4-ft. by 8-ft. by ½-in. sheets. COMFORTBASE is also available as a 4-ft. by 4-ft. by ½-in. sheet with scoring to conform to irregular surfaces. Treated to protect against termites, rot, and fungi. No urea formaldehyde or asbestos.

## 127 Homasote Co.

PO Box 7240, West Trenton, NJ 08628-0240
Phone: 800 257 9491 or 609 883 3300
Fax: 609 530 1584

**Price Index Number:** **0.96-1.3, 4-WAY/**
**1.9, 440**
**Price Index Standard:** **Douglas fir decking/**
**particleboard**

**CELLULOSE-FIBER DECKING**
EASY-PLY roof decking is tongue and groove, intended for exposed applications, fastened directly to rafters. Comes in 2-ft. by 8-ft. sheets. HOMASOTE 4-WAY floor decking comes in 2-ft. by 4-ft. by 8-ft. sheets 1¾ in. or 1¹¹⁄₃₂ in. thick. 4-WAY floor decking is a structural subfloor and carpet underlayment. No urea formaldehyde or asbestos additives. Class C flame-spread rating. 440 HOMASOTE fiberboard used under exterior siding or soffits, and ceilings. Comes in 4-ft. by 8-ft. sheets ½ in. to ⅝ in. thick.

## 128 Louisiana-Pacific

111 S.W. Fifth Ave., Portland, OR 97204
Phone: 800 999 9105 or 800 648 6895
Fax: 503 796 0204

**Price Index Number:** **0.90-1.10**
**Price Index Standard:** **plywood**

**LAMINATED SHEATHING**
OSB INNER SEAL can be used as wall or roof sheathing and tongue-and-groove flooring. Available in standard sheathing dimensions. OSB is dimensionally stable and durable with added rigidity over plywood for exterior walls. Company claims it uses a formaldehyde-free adhesive.

## 129 Oregon Strand Board

34363 Lake Creek Dr., Brownsville, OR 97327
Phone: 800 533 3374 or 541 466 5177
Fax: 541 466 5559

**Price Index Number:** **1.10**
**Price Index Standard:** **plywood**

**FLOOR AND WALL SHEATHING**
COMPLY composite plywood sheathing consists of 3 veneers of Douglas fir and 2 thick inner layers of reconstituted wood fiber bonded together. Cores allow use of non-veneer grade wood. Standard 4-ft. by 8-ft. size, 7 thicknesses, tongue-and-groove or square edge. Used for walls, floors (STURDI-FLOOR) and roofs although most applications are for subfloors. Exterior phenolic resin adhesives, moisture resistant, and guaranteed not to delaminate. COMPLY is a top-value solid core product with excellent fastener holding strength.

## 130 Simplex Products Division

PO Box 10, Adrian, MI 49221-0010
Phone: 800 345 8881    Fax: 517 265 7242

**Price Index Number:**    **0.70-0.90**
**Price Index Standard:**    **plywood**

### LAMINATED SHEATHING

THERMO-PLY insulative roof sheathing and siding are made from 100% recycled paper applied in plies. Roof sheathing comes with white polyethylene coating bonded to one side, and aluminum foil to the other. Water-resistant, recommended for use under tile roofs. R-value is R-.20. Comes in 4 ft. by 8 ft. by 0.078 in., 0.113 in. or 0.137 in. Wall sheathing can serve as a substitute for plywood depending upon structural requirements. Wall sheathing is available in the additional sizes of 48 in. or 48¾ in. by 9 ft. or 10 ft. Can be cut with a knife.

## 131 Tectum, Inc.

PO Box 3002, Newark, OH 43058
Phone: 614 345 9691
Fax: 800 832 8869 or 614 349 9305

**Price Index Number:**    **5.0-13.0**
**Price Index Standard:**    **plywood, ½ in.**

### CELLULOSE-FIBER DECKING

TECTUM is a structural roof panel manufactured without formaldehyde from aspen wood fibers with a binder of magnesium oxide extracted from seawater. Applicable to roofs and available in thicknesses from 1½ to 3 in. Other variations on this product include bonding TECTUM with insulation board and OSB. Aged R-values vary from R-4 to R-30 for an 8-in.-thick composite EPS panel.

## 132 Weyerhaeuser

4111 E. Four Mile Rd., Grayling, MI 49738
Phone: 517 348 2881 or 800 463 9378
(West Coast mill)
Fax: 517 348 8226

**Price Index Number:**    **0.90-1.10**
**Price Index Standard:**    **plywood**

### LAMINATED SHEATHING

STRUCTUREWOOD is an OSB product made from small, fast-growing trees. Available only on the East Coast. STURDIWOOD is the West Coast name for the above product made from flakes of aspen, poplar, and birch mixed with mostly isocyanate resins and wax with only trace amounts of formaldehyde. This product is dimensionally stable and durable with added rigidity over plywood for exterior walls. A typical ⁷⁄₁₆-in. OSB product starts as 4 plies 4 in. thick before being run through a press. Available in standard plywood dimensions.

**PREFABRICATED STRUCTURAL WOOD**      06170

## 133 Lamwood Systems, Inc.

1736 Boulder St., Denver, CO 80211
Phone: 800 826 8488 or 303 458 1736
Fax: 303 458 1739

**Price Index Number:**    **1.10, glu-lams**
**Price Index Standard:**    **2x12s and 2x10s,
construction-grade
Douglas fir**

### GLU-LAM BEAM

Manufacturer of glue-laminated beams and other structural wood products.

## 134 Louisiana-Pacific

N. 13455 Government Way,
Hayden Lake, ID 83835
Phone: 800 635 3437 or 208 772 6011
Fax: 208 772 7242

**Price Index Number:** 1.0+
**Price Index Standard:** wood stud

### FINGER-JOINTED STUDS
Finger-jointed studs manufactured from discarded kiln-dried 2x4s are straighter and dimensionally more stable than standard wood studs, reducing job-site waste. Material is recycled from standard wood stud mill. Vertical use only, for same structural performance as standard studs.

## 135 Timberweld

1643 24th St. W., Suite 308, Billings, MT 59102
Phone: 800 548 7069 or 406 652 3600
Fax: 406 652 3668

**Price Index Number:** 1.0
**Price Index Standard:** conventional wood-
framed vaulted roof

### GLU-LAM BEAM
Custom manufactured glued-laminated structural lumber.

## 136 Trus Joist MacMillan Paralam Div.

10277 154th St., Surrey, BC, Canada V3R 4J7
Phone: 604 588 7878   Fax: 604 589 9330

**Price Index Number:** 2.7
**Price Index Standard:** 2x12s, construction-
grade Douglas fir

### GLU-LAM CONSTRUCTION
PARALLAM parallel-strand lumber (PSL) from Douglas fir veneer.

## 137 Rosoboro Lumber Co.

PO Box 20, Springfield, OR 97477
Phone: 541 746 8411   Fax: 541 726 8919

**Price Index Number:** 1.05
**Price Index Standard:** standard glu-lam
beam

### GLU-LAM BEAM
REDI-LAM II beam is a combined glu-lam beam on top with a laminated-veneer lumber (LVL) beam on the bottom to create a lighter and less expensive beam than an LVL beam, but with equal strength. Has about a 20% greater allowable fiberstress than standard glu-lam, resulting in greater span and less deflection. Available in standard widths of 1¾ in., 3½ in., 5½ in., or 7 in. with camber as an option. Wood is Douglas fir.

## 138 Standard Structures, Inc.

340 Standard Ave., Windsor, CA 95492
Phone: 800 862 4936 or 707 544 2982
Fax: 707 544 2994

**Price Index Number:** 1.6
**Price Index Standard:** 2x12s, construction-
grade Douglas fir

### I-JOISTS AND GLU-LAMS
SSI joists with stress-rated flanges and OSB web. Finger-jointed and laminated framing lumber. Specialize in glued-laminated arches.

## 139 Trus Joist MacMillan

9777 W. Chinden Blvd., PO Box 60,
Boise, ID 83707
Phone: 800 628 3997 or 208 375 4754
Fax: 208 364 1300

**Price Index Number:** 1.10, SILENT FLOOR/
0.80, rim joist
**Price Index Standard:** 2x12s and 2x10s,
construction-grade
Douglas fir

**GLU-LAM BEAM**
Engineered wood products manufacturer whose products include PARALLAM beams, columns, and headers from parallel-strand lumber; MICRO=LAM laminated veneer lumber; SILENT FLOOR I-joists made from MICRO=LAM flanges and oriented strand board webs; and TIMBERSTRAND laminated-strand lumber. MICRO=LAM is for beams and headers, I-joists are for floor and roof structures, and TIMBERSTRAND is for light-duty headers and rim joists. The TIMBERSTRAND rim joists are 1¼ in. thick by 11⅞ in. wide by 11 ft. 8 in. or 17 ft. 6 in. long.

## 140 Unadilla Laminated Products

32 Clifton St., Unadilla, NY 13849
Phone: 607 369 9341   Fax: 607 369 3608

**Price Index Number:** 1.30
**Price Index Standard:** 2x12s, construction-
grade Douglas fir

**GLU-LAMS AND COLUMNS**
Glued-laminated columns, arches, trusses, and beams for both commercial and residential applications. Manufactured from southern pine and Douglas fir.

## 141 Willamette Industries Engineered Wood Products

PO Box 277, 1 E. Saginaw Rd., Saginaw, OR 97472
Phone: 541 942 4473 or 541 981 6003
Fax: 541 744 4653

**Price Index Number:** 0.92, STRUC-JOIST/
2.0, GLU-LAM
**Price Index Standard:** 2x12s and 4x12s,
construction-grade
Douglas fir

**I-JOISTS, GLU-LAMS, AND COLUMNS**
STRUC-JOISTS, for residential applications, and WS I-JOISTS, for commercial applications, are made from machine stress-rated (MSR) lumber flanges joined to OSB webs using a patented finger-joint (WSI). Joists are available in depths from 8 in. to 32 in. in 2-in. increments. Can be ordered with a taper. GLU-LAMINATED beams, columns, and headers are also available. Glu-lams come in standard sizes up to 130 ft. in length, in 3 finish grades. Can be ordered with camber. Fire rating is 1 hour. Manufactured from Douglas fir, larch, or southern pine.

## 142 Boise Cascade Corp.

PO Box 2400, White City, OR 97503-0400
Phone: 800 232 0788, 800 448 5602
or 541 826 1470   Fax: 541 826 0219

**Price Index Number:** 1.57/1.37
**Price Index Standard:** 2x10s and 2x12s,
construction-grade
Douglas fir

**I-JOISTS AND LVLS**
BCI I-joists are manufactured from plywood webs joining laminated-veneer lumber flanges. VERSA-LAM is a laminated-veneer lumber 1¾ in. thick, and 5½ in. to 14 in. deep. VERSA-LAM PLUS is 3½ in. and 5½ in. thick, and 5½ in. to 24 in. deep. I-joists come in 2 flange widths of 1¾ in. or 2⁵⁄₁₆ in., and in 5 depths from 9½ in. to 20 in. Available in lengths up to 66 ft. I-joist webs are detailed with finger joints for greater shear strength over butt joints.

## 143 Georgia-Pacific Corp.

360 Inverness Dr. S., Englewood, CO 80112
Phone: 800 423 2408   Fax: 904 680 1519

**Price Index Number:**   **1.9, GP-LAM 3⅛x12,**
     **for long spans**
**Price Index Standard:**   **4x12s, Douglas fir**

### I-JOISTS AND LVLS

GP-LAM laminated-veneer lumber (LVL) beams and headers for beams, headers, doors, windows, stairs, and furniture. WOOD I-BEAM with ⅜-in. OSB web and 2x4 flanges for roof and floor structures up to 48 ft. in length. WOOD I-BEAMS come in heights of 9½ in., 11⅞ in., 14 in., and 16 in. Phenolic resin adhesives.

## 144 Jager Industries, Inc.

8835 McLeod Trail S.W.,
Calgary, AB, Canada T2H OM3
Phone: 403 259 0700   Fax: 403 255 6008

**Price Index Number:**   **1.10**
**Price Index Standard:**   **2x12s, Douglas fir**

### I-JOISTS

TTS WOOD I-JOIST and SUPER JOIST from OSB between finger-jointed flanges for floor and roof joists. Joists are manufactured from spruce, pine and fir in standard and custom lengths up to 52 ft. TTS WOOD I-JOISTS have 2x4 flanges and come in depths from 9½ in. to 24 in. SUPER JOISTS have 2x3 flanges and come in depths of 9½ in. and 11 ½ in. Made from kiln-dried material.

## 145 Louisiana-Pacific

111 S.W. 5th Ave., Portland, OR 97204
Phone: 800 999 9105 or 503 221 0800
Fax: 503 796 0204

**Price Index Number:**   **1.06, TJI-15**
**Price Index Standard:**   **select structural**
     **grade 2x12**

### I-JOISTS AND LVLS

INNER-SEAL I-JOISTS, OSB web with 2x3 flange. GANG-LAM is laminated-veneer lumber for floor and roof structures. GNI joists with GANG-LAM flanges and OSB webs are available in 1½ in. or 1¾ in. standard or custom thicknesses, and depths from 9¼ in. to 24 in. All joists available in lengths up to 80 ft. Wood used in joists is southern pine for flanges and aspen or southern pine for OSB webs. GANG-LAM uses phenol formaldehyde, and GNI joists use phenol resorcinol adhesives.

## 146 Superior Wood Systems

PO Box 1208, Superior, WI 54880
Phone: 715 392 1822   Fax: 715 392 3484

**Price Index Number:**   **1.0-1.5**
**Price Index Standard:**   **2x12 header,**
     **Douglas fir**

### I-JOISTS

SWI-JOISTS with OSB webs and lumber flanges. SWII-HEADERS are insulated with polystyrene between two OSB webs and lumber flanges. Joists are for floor and roof structures. Headers replace sawn lumber and have an R-value of R-18. Joists are available in three flange sizes: 2 in. by 2 in., 2 in. by 3 in., or 2 in. by 4 in. Joist depths from 9¼ in. to 20 in. in lengths up to 48 ft. Headers have 2-in. by 4-in. or 2-in. by 6-in. flanges and depths of 7¼ in., 9¼ in., or 11¼ in.

## 147 Truswal

1101 N. Great Southwest Parkway,
Arlington, TX 76011
Phone: 800 521 9790   Fax: 817 652 3079

**Price Index Number:**   **1.10-1.12**
**Price Index Standard:**   **TJI with plywood web**

### TRUSSES

SPACEJOIST parallel-chord trusses featuring deep-V galvanized steel webs for high strength. SPACEJOISTS are used as rafters, joists, and studs where the open-web design serves for the placement of insulation, electrical, heating, and plumbing. Manufactured from Douglas fir or southern pine in depths from 9¼ in. to 19¾ in. with lengths up to 40 ft. Manufactured with 65% recycled steel in webs. Can be ordered with finger-jointed construction. Has a 1-hour fire rating.

*See also product numbers **289**, Georgia-Pacific Corp. (fiber-cement panels); and **305**, Phenix Biocomposites, Inc. (soy/resin composite).*

## 148 Advanced Environmental Recycling Technologies, Inc.

HC 10 Box 116, Junction, TX 76849
Phone: 800 951 5117 or 915 446 3430
Fax: 915 446 3864

**Price Index Number:** 1.0
**Price Index Standard:** conventional milled wood products

### WOOD/RESIN COMPOSITE

Manufactures composite materials, such as MOISTURESHIELD, from recycled wood fiber and polyethylene that are resistant to rotting, warping, cracking, and dimensional instability. Recycled wood fiber is mostly cedar. Used by window and door manufacturers as a substrate for such items as door rails and window sills. Made to custom order and shipped by the truckload to manufacturers.

## 149 Architectural Forest Enterprises

1030 Quesada Ave., San Francisco, CA 94124
Phone: 800 483 6337 or 415 622 7300
Fax: 415 822 8540

**Price Index Number:** 1.05
**Price Index Standard:** plywood, A-face

### FIBERBOARD

ECO-PANEL manufactured from MEDITE II medium-density fiberboard, recycled paper, and wood veneers such as maple and cherry. Combines high-quality formaldehyde-free product with sustainable-yield forest sources. Used in furniture, cabinets, and partitions. Available in thicknesses from ¼ in. to 1½ in.

## 150 BioFab

PO Box 990556, Redding, CA 96099
Phone: 916 243 4032    Fax: 916 243 4032

**Price Index Number:** 0.7 PGB material + labor
**Price Index Standard:** 2x4 framed wall

### PARTICLEBOARD

PACIFIC GOLD BOARD (PGB) is a straw panel compressed and bonded with heat and covered with 85 lb. recycled kraft paper. Available in 2¼-in. by 4-ft. by 8-ft. panels to replace non-load-bearing walls. Comes with electrical chases precut into each panel. Joints can be taped and floated like conventional gypsum-board walls. PACIFIC GOLD BOARD BALED BATT (PGB3) is a straw bale woven with wire 1½ in. by 4 ft. by 8 ft. for a thatched ceiling finished surface. Absorbs sound well. Can be cut with a saw.

## 151 Contact Lumber

1881 S.W. Front Ave., Portland, OR 97201
Phone: 800 547 1038 or 503 228 7361
Fax: 503 221 1340

**Price Index Number:** 0.90-1.05
**Price Index Standard:** solid oak or maple

### TRIM PRODUCTS

Jamb, molding, and other trim products made from thin veneers of finish wood laminated over a core of finger-jointed lumber in 14-ft. lengths. Most common veneers are oak and maple. The finger-jointed core avoids warping or twisting and is an efficient use of wood fiber.

## 152 Evanite Fiber Corp.

PO Box E, Corvallis, OR 97333
Phone: 541 753 1211    Fax: 541 753 0336

**Price Index Number:** 4.5
**Price Index Standard:** Masonite hardboard, ¼ in. thick

### FIBERBOARD

EVANITE FIBER hardboard manufactured from waste wood, such as pallets, crates, and construction debris, which is chipped, mixed with sawdust and plywood manufacturing waste, and heat-pressurized to form a solid panel. For furniture and cabinet backing and drawer bottoms. Comes in ⅛-in. or ¼-in. by 4-ft. by 8-ft. sheets.

## 153 Formica Corp.

10155 Reading Rd., Cincinnati, OH 45241
Phone: 800 367 6422 or 513 786 3400
Fax: 513 786 3542

**Price Index Number:** **1.2**
**Price Index Standard:** **quartersawn ash**
**veneer**

### WOOD VENEER
LIGNA wood veneers are made from sustainably harvested Italian poplar to look like the wood grains of tropical hardwoods. Logs are reconstructed from Italian poplar to create the desired species wood-grain appearance in the veneer.

## 154 Gridcore Systems

1400 Canal Ave., Long Beach, CA 90813
Phone: 310 901 1492   Fax: 310 901 1499

**Price Index Number:** **2.4**
**Price Index Standard:** **particleboard**

### CELLULOSE PANEL
GRIDCORE is a nonstructural panel with a high strength-to-weight ratio manufactured from 100% recycled paper without toxic additives. Product is one-third the weight of particleboard at less than 1 lb./sq. ft. Available in 4-ft. by 10-ft. panels and worked with standard hand tools as a substrate for partitions, furniture, counters and other nonstructural appllications. Must be sealed from moisture. Class C fire rating. A structural panel is being developed. Manufacturer claims this product to be nontoxic.

## 155 Homasote Co.

PO Box 7240, West Trenton, NJ 08628-0240
Phone: 800 257 9491 or 609 883 3300
Fax: 609 530 1584

**Price Index Number:** **0.70, NOVA/1.9, 440/**
**0.96-1.3, 4-WAY**
**Price Index Standard:** **cork board/**
**particleboard/tongue-**
**and-groove decking**

### FIBERBOARD
NOVA CORK interior panels of natural, virgin cork laminated to HOMASOTE. 440 HOMASOTE fiberboard is used under exterior siding or soffits or in ceilings. HOMASOTE 4-WAY floor decking comes in 2-ft. by 8-ft. or 4-ft. by 8-ft. sheets $1\frac{3}{4}$ in. or $1\frac{11}{32}$ in. thick, for 24 in. o.c. or 16 in. o.c., respectively, combining insulation with acoustical properties. Treated to protect against termites, rot, and fungi. No urea formaldehyde or asbestos.

## 156 J Squared Timberworks, Inc.

449 N. 34th St., Seattle, WA 98103
Phone: 800 598 3074 or 206 633 0504
Fax: 206 633 0565

**Price Index Number:** **1.20-1.25**
**Price Index Standard:** **millwork from**
**virgin wood**

### RECYCLED LUMBER
Architectural millwork custom made from remilled salvaged lumber. Products include doors, staircases, wainscots, and railings. Products are typically made from a very stable, air-dried, tight-grain wood, which may or may not have an antiqued appearance. J Squared also provides a design and construction service for timber-framed houses.

## 157 Meadowood Industries, Inc.

33242 Red Bridge Rd., Albany, OR 97321
Phone: 541 259 1303   Fax: 541 259 1355

**Price Index Number:** **1.0**
**Price Index Standard:** **plywood**

### PARTICLEBOARD
Compressed straw board pressed together with phenolic resins into a composite panel using agricultural by-products. Used as a replacement for nonstructural particleboard for furniture and kitchen cabinets. Comes in ¼-in. by 4-ft. by 8-ft. sheets. The company will work with clients to create custom patterns made from ferns and dried flowers pressed into the surface. No formaldehyde in the product.

## 158 Medite Corp.

PO Box 4040, Medford, OR 97501
Phone: 800 676 3339 or 541 773 2522
Fax: 541 779 9921

**Price Index Number:** **1.0, MEDITE II/**
**1.10, MEDEX**
**Price Index Standard:** **maple veneer plywood**

### FIBERBOARD
Similar to medium-density fiberboard (MDF), MEDEX is an exterior-grade board produced with a polyurea resin matrix adhesive instead of urea-formaldehyde or phenol formaldehyde. Made from waste wood fiber from milling operations. MEDITE II is a new formaldehyde-free board produced as the original MEDEX, but at lower cost for interior applications in dry areas. Available in sizes up to 5 ft. by 18 ft. by 1¼ in. thick. MEDITE-FR is a Class A fire-retardant board that is formaldehyde free.

## 159 PrimeBoard, Inc.

2111 North 3M Dr., Wahpeton, ND 58075
Phone: 800 943 2823 or 701 642 1152
Fax: 701 642 1154

**Price Index Number:** **1.25-1.3**
**Price Index Standard:** **plywood**

### PARTICLEBOARD
Compressed wheat straw fiberboard, steam heated and pressed together into a composite panel using agricultural by-products such as wheat, is used as a replacement for nonstructural particleboard. Used in kitchen cabinets and countertops. 10% lighter in weight than particleboard. Comes in ⅜-in. and ¾-in. thicknesses, in 5-ft. by 8-ft., 5-ft. by 9-ft., or 5-ft. by 10-ft. sheets. Meets commercial-grade M3 Standards. No formaldehyde in the product.

## 160 Barrier Technology USA

510 Fourth St. N, Watkins, MN 55389
Phone: 800 638 4570   Fax: 320 764 5799

**Price Index Number:** **2.4, BLAZEGUARD +**
**plywood**
**Price Index Standard:** **plywood**

### FIRE-RATED SHEATHING
BLAZEGUARD is a fire-rated roof and wall sheathing that exceeds 30-minute Class A flame-spread requirements. The fire shield is made of a natural crystalline-structured laminate containing molecularly bound water. The fire shield is factory applied to plywood or OSB substrates using proprietary adhesive systems. Can be cut, drilled, nailed, and stapled using common carpentry tools. No special handling required. Manufacturer claims this product to be nontoxic.

## 161 Chemical Specialties, Inc.

1 Woodlawn Green, Suite 250, Charlotte, NC 28217
Phone: 800 421 8661 or 704 522 0825,
Fax: 704 527 8232

**Price Index Number:** **2.6**
**Price Index Standard:** **chromated copper**
**arsenate (CCA)**

### PRESERVATIVE
ACQ PRESERVE is used for pressure treating wood for exterior applications. Classified as a nonhazardous chemical.

## 162 Eco Design/Natural Choice

1365 Rufina Circle, Santa Fe, NM 87505
Phone: 800 621 2591 or 505 438 3448
Fax: 505 438 0199

**Price Index Number:** **3.9**
**Price Index Standard:** **copper napthenate**
**with VOCs**

### PRESERVATIVE
LIVOS DONNOS PRESERVATIVE is a spray- or brush-applied wood-pitch product. First-coat coverage is approximately 120 sq. ft. to 150 sq. ft. per liter, depending on surface absorption. The company is the only importer of Livos Plantchemistry products.

## 163 Nisus Corp

215 Dunarant Dr., Rockford, TN 37853
Phone: 800 264 0870 or 423 577 6119
Fax: 423 577 5825

**Price Index Number:** **3.0-4.0**
**Price Index Standard:** **chromated copper arsenate (CCA)**

### PRESERVATIVE
BORA-CARE is a boron wood preservative that uses disodium octaborate tetrahydrate (DOT) as its active ingredient and is considered a relatively nontoxic wood-preservative treatment. It is available in liquid form and is registered with the EPA as an insecticide. This product is actually U.S. Borax's TIM-BOR product that has been premixed with a glycol solution to promote diffusion into dry lumber.

## 164 Prime Line Decorating

4854 W. 61st Dr., Arvada, CO 80003
Phone: 888 966 3476 or 303 650 1681
Fax: 303 428 3004

**Price Index Number:** **2.8**
**Price Index Standard:** **copper napthenate with VOCs**

### PRESERVATIVE
WOOD IRON wood preservative is a mixture of all natural oils with low toxic additives to resist fungus and mildew growth. Intended for aboveground on-site brush-on applications.

## 165 Tallon Termite and Pest Control

30073 Ahern St., Union City, CA 94587
Phone: 800 809 2653   Fax: 510 429 6829

**Price Index Number:** **1.0**
**Price Index Standard:** **standard toxic poison methods**

### INSECT TREATMENT
Nontoxic wood boring insect treatment. Article in *Old House Journal,* March/April 1991, describes injection of liquid nitrogen and heat. Dry-wood termites are frozen with liquid nitrogen, and subterranean termites are attacked with nematodes. 2-year guarantee. No charge for a second opinion.

## 166 U.S. Borax, Inc.

26877 Tourney Rd., Valencia, CA 91355-1847
Phone: 800 984 6267 or 805 287 5400
Fax: 805 287 5495

**Price Index Number:** **1.0**
**Price Index Standard:** **chromated copper arsenate (CCA)**

### PRESERVATIVE
TIM-BOR is a boron wood preservative that uses disodium octaborate tetrahydrate (DOT) as its active ingredient and is considered relatively nontoxic. It is available in powder form only to licensed pest-control applicators because it is registered with the EPA as an insecticide. The powder is mixed with water to create a 10% to 20% solution for either spraying onto structures or dipping. TIM-BOR is for wet lumber (greater than 20% moisture content) that is not exposed to the weather or is not in contact with the ground because the product tends to diffuse out of wood under wet conditions. The manufacturer claims that treating the entire house with TIM-BOR during construction adds about a $2,000 cost to a 2,000-sq.-ft. residence.

## 167 Weatherall Northwest

658 Hwy 93 S., Hamilton, MT 59840
Phone: 800 531 2286 or 406 363 4262
Fax: 406 363 1558

**Price Index Number:** **1.9**
**Price Index Standard:** **copper napthenate with VOCs**

### PRESERVATIVE
Nontoxic, nonflammable acrylic latex wood preservative for exterior and interior. First coat coverage is 240 sq. ft. to 500 sq. ft. per gal.

### 168  Temple-Inland Forest Products

PO Drawer N, Diboll, TX 75941
Phone: 800 231 6060 or 409 829 5511
Fax: 800 426 7382

**Price Index Number:   1.34**
**Price Index Standard:   select heart redwood**

**FIBERBOARD**
TRIMCRAFT is a nonstructural, easy-to-work wood fiber and resin composite made from wood chips for exterior trim applications. Stable, factory-primed. Comes in 16-ft. length, ¾-in. thickness, and 4-in. to 12-in. widths. Comes with factory-applied oven-baked primer with a 5-year warranty.

### 169  Aeolian Enterprises

PO Box 888, Latrobe, PA 15661
Phone: 412 539 9460   Fax: 412 539 0572

**Price Index Number:   0.60-1.35**
**Price Index Standard:   select heart redwood**

**PLASTIC LUMBER**
Hollow plastic lumber manufactured from 100% recycled HDPE from industrial and post-consumer sources.

### 170  Alket Industries

PO Box 849, 2148 Hwy 22 W., Kalona, IA 52247
Phone: 319 656 5100   Fax: 319 656 5105

**Price Index Number:   2.0**
**Price Index Standard:   lumber**

**PLASTIC LUMBER**
Manufacturer of plastic lumber and sheeting. Also provides recycled high-density polyethylene laminated onto substrates, such as plywood or MDF, for applications such as public restroom walls, freezer liners, and garage and outbuilding walls. Excellent anti-graffiti surface; surface repairs easily.

### 171  Amazing Recycled Products

PO Box 312, Denver, CO 80201
Phone: 800 241 2174 or 303 699 7693
Fax: 303 778 9120

**Price Index Number:   1.1**
**Price Index Standard:   lumber**

**PLASTIC LUMBER**
Recycled plastic lumber with 5% to 10% cellulose added to create a dull-colored, nonslick finish.

### 172  Bedford Industries

1659 Rowe Ave., PO Box 39,
Worthington, MN 56187-0039
Phone: 800 533 5314 or 507 376 4136
Fax: 507 376 6742

**Price Index Number:   1.0-1.85**
**Price Index Standard:   select heart redwood**

**PLASTIC LUMBER**
Plastic timber manufactured from recycled HDPE from industrial sources plus a paper binder. Available in black, brown, gray, and cedar.

## 173 Carrysafe

920 Davis Rd., Elgin, IL 60123
Phone: 847 931 4528   Fax: 847 931 1771

**Price Index Number:   1.8-3.2**
**Price Index Standard:   select heart redwood**

**PLASTIC LUMBER**
CARRYSAFE RE-SOURCE LUMBER made from recycled HDPE derived from milk jugs. Can be nailed, sawn, and drilled. Product includes 1x6 tongue-and-groove decking to replace 2x6 redwood decking. Available in gray weathered redwood, cedar, and light oak colors. Black is also available and is the least expensive for utilitarian applications.

## 174 Cascades Re-Plast, Inc.

1350 Chemin Quatres Saisons
Notre-Dame-du-bon-Conseil, QC, Canada J0C 1A0
Phone: 819 336 2440   Fax: 819 336 2442

**Price Index Number:   1.6-2.3**
**Price Index Standard:   select heart redwood**

**PLASTIC LUMBER**
Plastic lumber and exterior furniture manufactured from 100% post-consumer thermoplastics from curbside collection sources.

## 175 Eagle One Golf Products

1201 W. Katella Ave., Orange, CA 92867
Phone: 800 448 4409 or 714 997 1400
Fax: 714 997 3400

**Price Index Number:   0.80**
**Price Index Standard:   virgin rubber mats**

**PLASTIC LUMBER**
100% recycled plastic lumber.

## 176 Eagle Recycled Products

1201 W. Katella Ave., Orange, CA 92867
Phone: 800 448 4409 or 714 997 1400
Fax: 714 997 3400

**Price Index Number:   3.1-4.9**
**Price Index Standard:   select heart redwood**

**PLASTIC LUMBER**
Plastic lumber and exterior furniture manufactured from 100% recycled plastic.

## 177 Eaglebrook Products, Inc.

2650 W. Roosevelt Rd., Chicago, IL 60608
Phone: 312 638 0033   Fax: 312 638 2567

**Price Index Number:   2.0, 2x4/1.15, 4x4**
**Price Index Standard:   select heart redwood**

**PLASTIC LUMBER**
DURAWOOD plastic lumber consists of HDPE with UV-inhibiting pigment systems. Claims more than 90% recycled post-consumer waste by weight. Makes landscape ties and site furnishings.

## 178 Earth Care Products of America

2300 W. Glade Rd., Suite 440 West
Boca Raton, FL 33431
Phone: 800 653 2784   Fax: 407 394 5335

**Price Index Number:** 1.2-1.7
**Price Index Standard:** select heart redwood

**PLASTIC LUMBER**
Plastic lumber manufactured from 100% recycled post-industrial and post-consumer commingled plastic waste. Available in all nominal sizes from 2x2 to 2x10 and 4x4 and 4x6. Also has 5/4 boards, car and truck stops, and site furnishings.

## 179 Environmental Specialty Products, Inc.

PO Box 1114, Guasti, CA 91743
Phone: 800 775 2784 or 909 390 8800
Fax: 909 390 8700

**Price Index Number:** 0.80, TREX/
1.6, plastic lumber
**Price Index Standard:** select heart redwood

**PLASTIC LUMBER**
Manufacturer and distributor of plastic lumber and furniture from recycled plastic. TREX is approximately 50% post-consumer thermoplastic and 50% recycled wood fiber. Good for decks.

## 180 Metro Plastics, Inc.

2916 107th St. S., Tacoma, WA 98444
Phone: 800 676 4091   Fax: 206 588 3039

**Price Index Number:** 0.82-1.2, DURABORD
**Price Index Standard:** select heart redwood

**PLASTIC LUMBER**
DURAPOST plastic fences and DURABORD plastic lumber manufactured from 100% post-consumer recycled plastic.

## 181 Obex, Inc.

PO Box 1253, Stamford, CT 06904
Phone: 800 876 8735 or 203 975 9094
Fax: 203 975 9403

**Price Index Number:** 4.3, garden tiles
**Price Index Standard:** interlocking concrete
landscape pavers

**PLASTIC LUMBER**
NOVAWOOD landscape ties, 4 in. by 5 in. by 8 ft., manufactured from recycled plastics from residential, commercial, and industrial plastic waste. GARDEN TILES, 24 in. by 24 in. by 3 in., are slate gray and can be arranged in a parquet pattern as a replacement for conventional patio pavers. The company also has a 27-cu.-ft. compost bin manufactured from recycled plastics and a raised-garden kit.

## 182 Phoenix Recycled Plastics

225 Washington St., Conshohocken, PA 19428
Phone: 610 940 1590   Fax: 610 940 1593

**Price Index Number:** 1.28-1.43
**Price Index Standard:** select heart redwood

**PLASTIC LUMBER**
Fiberglass-reinforced plastic lumber manufacturer claims its product can be used for joists, beams, substructure, and bulkheading. Also manufactures a line of outdoor products including decks, marina docks, landscape ties, car stops, speed bumps, picnic tables, benches, and jungle gyms made from mostly recycled milk jugs.

## 183 The Plastic Lumber Co.

540 S. Main St., Bldg. 7, Akron, OH 44311
Phone: 800 886 8990 or 330 762 8989
Fax: 330 762 1613

**Price Index Number:** **1.4**
**Price Index Standard:** **select heart redwood**

**PLASTIC LUMBER**
Broad supply of plastic lumber and plastic lumber products such as site furniture, car stops, and speed bumps from made from commingled recycled plastics. The source of plastics are mostly post-consumer milk jugs, bubble wrap, shrink wrap, and aseptic packaging.

## 184 Plastic Recycling, Inc.

10252 Highway 65 North,
Iowa Falls, IA 50126-8823
Phone: 800 338 1438 or 515 648 5073
Fax: 515 648 5074

**Price Index Number:** **1.0+**
**Price Index Standard:** **redwood, B-grade**

**PLASTIC LUMBER**
Plastic lumber for benches, tables, and fencing. Also a range of other products including speed bumps, car stops, and site furniture.

## 185 Recycled Plastic Man, Inc.

PO Box 3368, Venice, FL 34293
Phone: 941 497 1020    Fax: 941 473 0131

**Price Index Number:** **1.2-1.3**
**Price Index Standard:** **select heart redwood**

**PLASTIC LUMBER**
Plastic lumber, exterior furniture, car stops, parking blocks, and piling docks manufactured from 100% recycled plastic.

## 186 Recycled Plastics Industries, Inc.

1011 McDonald St., Green Bay, WI 54303
Phone: 414 433 0900    Fax: 414 433 9329

**Price Index Number:** **1.3-1.9**
**Price Index Standard:** **select heart redwood**

**PLASTIC LUMBER**
RECYCLEMAID plastic lumber manufactured from recycled milk jugs. Available in most nominal sizes from 1x4 up to 2x10 in black, gray, cedar, and white. Custom profiles also available with setup charge for small runs. Manufactures plastic benches and tables also.

## 187 Recycled Polymer Associates

152 W. 26th St., New York, NY 10001
Phone: 212 463 8622    Fax: 212 635 5777

**Price Index Number:** **0.85-1.4**
**Price Index Standard:** **select heart redwood**

**PLASTIC LUMBER**
Plastic lumber and hollow plastic fencing manufactured from 95% recycled polyolefins and 5% agents and inert materials.

## 188 Renewed Materials Industries

621 W. Division St., Muenster, TX 76252
Phone: 817 759 4181    Fax: 817 759 4011

**Price Index Number:**    **1.7**
**Price Index Standard:**    **construction-grade redwood**

### PLASTIC LUMBER

RUMBER 2x8 interlocking boards made from a combination of recycled plastic and tire rubber. The rubber and plastic combination gives this product a much lower coefficient of expansion than standard plastic lumber. The manufacturer claims RUMBER will expand only ⅜ in. over a 20-ft. length. This tongue-and-groove product is used for heavy-duty flooring and retaining walls.

## 189 Trex Wood Polymer

20 S. Cameron, Winchester, VA 22601
Phone: 800 846 2739 or 800 289 8739
Fax: 203 831 4222

**Price Index Number:**    **0.80, TREX**
**Price Index Standard:**    **select heart redwood**

### PLASTIC LUMBER

TREX is approximately 50% post consumer thermoplastic and 50% recycled wood fiber. TREX Recycling Program accepts TREX for reprocessing as long as it is free of coatings, nails, screws, etc.

## 190 Trimax Lumber

2076 5th Ave., Ronkonkoma, NY 11779
Phone: 516 471 7777    Fax: 516 471 7862

**Price Index Number:**    **1.4**
**Price Index Standard:**    **select heart redwood**

### PLASTIC LUMBER

TRIMAX is 100% curbside HDPE manufactured from recycled polyolefin waste with fiberglass added for strength. Manufacturer claims equal in strength to southern yellow pine No. 2, ASTM 198 testing standard. Can be worked with conventional tools and fasteners. Used by manufacturers of site furnishings.

## 191 Formica Corp.

10155 Reading Rd., Cincinnati, OH 45241
Phone: 800 367 6422 or 513 786 3400
Fax: 513 786 3542

**Price Index Number:**    **0.60**
**Price Index Standard:**    **Corian, installed**

### PLASTIC LAMINATE

FORMICA plastic laminates for countertops use wood pulp from recycled print and kraft paper. Layers of paper are pressed together with resins to make the substrate. The finished surface is created by pressing a melamine layer with a stainless steel plate. This process replaces the old method of using aluminum foil and paper only once before it is discarded. Some manufacturers of similar plastic laminates continue to utilize the old process.

## 192 Global Plastic Products, Inc.

3400 Peachtree Rd., Atlanta, GA 30326
Phone: 404 239 6277    Fax: 404 239 6284

**Price Index Number:**    **3.0, PLASALLOY**
**Price Index Standard:**    **plywood**

### SOLID SHEET PLASTIC

INNOPLAST GP sheets are manufactured from recycled plastics consisting of 50% thermoplastic polyethylene with various other plastics and impurities embedded. Available in a standard size of 40 in. by 48 in., in thicknesses of ¼ in., ⅜ in., ½ in., and up to 1½ in. PLASALLOY is a commodity-grade, high-performance, waterproof composite of hemp-like natural fibers and HDPE or HDPP extruded or compression molded to specifications and manufactured from recycled materials upon request.

### 193 Santana Products, Inc.

PO Box 2021, Scranton, PA 18501
Phone: 800 233 4701 or 717 343 7921
Fax: 717 348 2959

**Price Index Number:** 1.0
**Price Index Standard:** virgin material

**SOLID SHEET PLASTIC**
1-in.-thick laminated panels of plastic manufactured from HDPE. Virgin plastic caps are laminated to a recycled plastic core in a heat-molding process to produce one homogeneous panel. Total recycled material is a minimum of 10%, depending upon availability. Used in public restrooms for toilet compartments. Product can also be used in shower compartments and lavatory countertops. It is moisture resistant, and little maintenance is required. Available in various sizes and colors including a marble finish.

### 194 Spartech Plastics

PO Box 757, Richmond, IN 47375
Phone: 800 428 6573 or 317 935 7541
Fax: 317 935 4685

**Price Index Number:** 0.90
**Price Index Standard:** virgin material

**SOLID SHEET PLASTIC**
Sheet plastic and roll stock manufactured from various types of recycled plastics. Typical customers include thermal formers manufacturing various plastic products such as truck-bed liners. Will also custom fabricate products.

### 195 Yemm and Hart Green Materials

RR 1, Box 173, Marquand, MO 63655-9610
Phone: 573 783 5434   Fax: 573 783 5434

**Price Index Number:** 0.85, plastic panels
**Price Index Standard:** virgin material

**PLASTIC PANELS**
Plastic panels, sheeting, and furniture from recycled plastic sources. A variety of plastic products are available, not all of which are suitable for building and landscaping applications.

### 196 AFM Enterprises, Inc.

350 W. Ash St., Suite 700, San Diego, CA 92101
Phone: 619 239 0321   Fax: 619 239 0565

**Price Index Number:** 1.10
**Price Index Standard:** standard VOC sealers

**WATERPROOFING**
Paints for sealing concrete, masonry, wood, and metal. Also available are clear sealers, stains, and cleaning products for surfaces including carpets. These products use the Huls Coloring System and are specially designed for people sensitive to interior air quality and chemicals.

### 197 Mameco/Paramount Technical Products

4475 E. 175th St., Cleveland, OH 44128
Phone: 216 752 4400, 800 658 5500
or 605 642 4787   Fax: 216 752 5005

**Price Index Number:** 1.0
**Price Index Standard:** rubber sheeting + labor

**MEMBRANE WATERPROOFING**
PARASEAL is a waterproofing sheet made from a 20-mil or 40-mil HDPE membrane covered on one side with 1 lb./sq. ft. of expandable bentonite clay. This is a self-seaming and self-healing waterproofing product that seals tightly to itself and any other surface when in the presence of water. Sheets are available in 4-ft. by 24-ft., 8-ft. by 200-ft., or 17½-ft. by 200-ft. rolls. Company manufactures a complete line of bentonite waterproofing products for various applications.

## 198 Rubber Polymer Corp.

1135 W. Portage Trail, Akron, OH 44313
Phone: 800 860 7721   Fax: 330 445 9416

**Price Index Number:** 1.6-2.0/1.2-5.0
**Price Index Standard:** Bituthene/asphalt-based coating

### FLUID-APPLIED WATERPROOFING

RUB-R-WALL foundation and basement waterproofing made from a proprietary mixture of several synthetic rubber co-polymers. Requires a 2-ft. foundation backcut for spray application at high pressure (3000 psi) and high temperature (approximately 150°F). Protection board is installed. Residential applications carry a limited lifetime warranty, which is transferable to new owners. Commercial applications have a 10-year warranty. GRAYWALL is a synthetic rubber dampproofing used where an asphalt-based coating might typically be applied. Manufacturer claims products meet the strictest VOC environmental regulations and do not have any toxins that could leach into surrounding groundwater.

## 199 Xypex Chemical Corp.

13731 Mayfield Pl., Richmond, BC, Canada V6V 2G9
Phone: 604 273 5265   Fax: 604 270 0451

**Price Index Number:** 1.0
**Price Index Standard:** rubber sheeting + labor

### CRYSTALLINE BARRIER WATERPROOFING

XYPEX is a crystalline barrier waterproofing, which when combined with or applied to concrete forms crystals in the presence of water to seal leaks. These crystals continue to grow within the concrete as long as there is water and calcium hydroxide present. This process narrows the capillary tracks until the passage of water is blocked, stopping the crystal growth. Manufacturer claims this product to be nontoxic.

## 200 Hydrozo, Inc.

8570 Philips Highway, Suite 108,
Jacksonville, FL 32256
Phone: 800 422 1902 or 904 828 4900
Fax: 904 828 4992

**Price Index Number:** 0.87
**Price Index Standard:** solvent-based sealants

### FLUID-APPLIED WATER REPELLENT

ENVIROSEAL DOUBLE 7 FOR BRICK and EVIROSEAL 20 FOR CONCRETE penetrating sealers for repelling water and chloride ions from above-grade concrete, brick, and stone surfaces. Increases life of surfaces and protects rebars from rusting. Typically reapplied every 5 years. Meets strictest VOC environmental regulations.

## 201 W. R. Meadows

PO Box 543, Elgin, IL 60121
Phone: 800 342 5976 or 847 683 4500
Fax: 847 683 4544

**Price Index Number:** 0.85-0.90
**Price Index Standard:** solvent-based sealants

### FLUID-APPLIED WATER REPELLENT

VOCOMP-20, VOCOMP-25, VOCOMP-30, and VOCOMP-35 curing and sealing compounds for concrete. Meets strictest VOC environmental regulations.

## 202 Resource Conservation Tech.

2633 N. Calvert St., Baltimore, MD 21218
Phone: 410 366 1146   Fax: 410 366 1202

**Price Index Number:** **25.0, HYGRODIODE/**
**4.0, TENO-ARM**
**Price Index Standard:** **visqueen**

**VAPOR RETARDER**
HYGRODIODE is a one-way vapor barrier for use where there is too little room for roof-cavity ventilation. TENO-ARM is a polyethylene vapor barrier designed to last for hundreds of years. Manufacturer also makes wiring box enclosures, plastic boxes to enclose and seal electric outlets, and cover plates that automatically seal when no plug is in.

## 203 Advanced Foam Plastics, Inc.

5250 N. Sherman St., Denver, CO 80216
Phone: 800 525 8697 or 303 297 3844
Fax: 303 292 2613

**Price Index Number:** **0.95**
**Price Index Standard:** **virgin expanded polystyrene**

**EXPANDED POLYSTYRENE**
Expanded polystyrene insulation manufactured with 10% recycled material, without CFCs, HCFCs, or formaldehyde. Uses the inert gas pentane as a blowing agent, which does not destroy atmospheric ozone or contribute to global warming. A 1-lb. density has an R-value of R-3.1/in. and a 2-lb. density is R-4.35/in. Manufacturer offers a 20-year guarantee on insulation value. Just opened a new plant in Reno, Nev.

## 204 All-Weather Insulation Co.

5309 Bardstown, Springfield, KY 40069
Phone: 800 467 0211 or 606 336 3931
Fax: 606 336 9289

**Price Index Number:** **0.90**
**Price Index Standard:** **fiberglass batt**

**LOOSE-FILL INSULATION**
Cellulose insulation made from recycled newspapers with boric acid added as a fire retardant.

## 205 American Insulation, Inc.

600 Industrial Dr., Bloomer, WI 54724
Phone: 800 633 3179 or 715 568 3898
Fax: 715 568 3897

**Price Index Number:** **0.90**
**Price Index Standard:** **fiberglass batt**

**LOOSE-FILL INSULATION**
Cellulose insulation made from 100% recycled newsprint with soybean-based inks. Fire retardant formula is proprietary. Distribution limited to the Midwest.

## 206 American Rockwool, Inc.

PO Box C, 1 Jackrabbit Rd., Nolanville, TX 76559
Phone: 800 792 3539 or 817 698 2233
Fax: 817 698 2234

**Price Index Number:** **0.7 material/1.30**
**material + labor**
**Price Index Standard:** **fiberglass batt**

**MINERAL-WOOL INSULATION**
Mineral-wool insulation manufactured from steel slag with a 75% recycled content. Inert and nonflammable. Wall R-value is R-4.13/in. Attic R-value is R-2.97/in. Contains no asbestos or phenols. Will not settle, rot, decay, or breakdown.

## 207  American Sprayed Fibers, Inc.

1550 E. 91st Dr., Merrillville, IN 46410
Phone: 800 824 2997 or 219 769 0180
Fax: 219 736 6126

**Price Index Number:**  **0.5**
**Price Index Standard:  fiberglass batt**

**FIBER INSULATION**
SOUND-PRUF spray-on acoustical insulation manufactured from rock wool, cellulose, and a liquid adhesive. R-value is R-3/in. to R-3.5/in. Manufacturer claims product is nontoxic and does not support the growth of bacteria or fungus.

---

## 208  Applegate Manufacturing

1000 Highview Dr., Webberville, MI 48892
Phone: 800 627 7536 or 517 521 3545
Fax: 517 521 3597

**Price Index Number:**  **0.63 material/1.30**
                        **material + labor**
**Price Index Standard:  fiberglass batt**

**LOOSE-FILL INSULATION**
Cellulose insulation made from recycled newsprint with boric acid added as a fire retardant.

---

## 209  Applegate Manufacturing

1850 1B Maugans Ave., Hagerstown, MD 21742
Phone: 800 231 1939 or 301 791 7360
Fax: 301 791 2143

**Price Index Number:**  **0.63 material/1.30**
                        **material + labor**
**Price Index Standard:  fiberglass batt**

**LOOSE-FILL INSULATION**
Cellulose insulation made from recycled newsprint with boric acid added as a fire retardant.

---

## 210  Arctic Manufacturing

1216 N. 11th Ave., Greeley, CO 80631
Phone: 970 353 2392   Fax: 970 304 0806

**Price Index Number:**  **0.63 material/1.30**
                        **material + labor**
**Price Index Standard:  fiberglass batt**

**LOOSE-FILL INSULATION**
CLEAN cellulose insulation made from recycled newsprint with boric acid added as a fire retardant. R-value of R-3.7/in. or R-13.25 for 4 inches. Made from recycled newsprint, manufacturer claims CLEAN will not settle.

## 211 Ark-Seal, Inc. International

2120 S. Kalamath, Denver, CO 80223
Phone: 800 525 8992   Fax: 303 934 5240

**Price Index Number:   1.33**
**Price Index Standard:   fiberglass batt**

### SPRAYED INSULATION

BLOW-IN BLANKET SYSTEM (BIBS) combines fiberglass or cellulose with a latex binder to prevent settling. Product is blown into wall or ceiling cavity or onto netting. Tested R-value is R-4.18/in. for fiberglass and R-3.64/in. to R-3.82/in. for cellulose. Excellent acoustical rating because all voids are filled. FIBERIFFIC is a commercial insulation system combining various loose-fill insulations with a latex binder. FIBERIFFIC is a more expensive application of the BIBS, which is currently only marketed and available outside of the United States. Although the manufacturer claims no outgassing, paint smell after installation similar to a freshly painted space could be a problem for chemically sensitive people.

## 212 BMCA Insulation Products, Inc.

300 N. Haven Ave., Ontario, CA 91761
Phone: 800 858 8868 or 909 390 8811
Fax: 909 390 8764

**Price Index Number:   2.0**
**Price Index Standard:   polyiso per R-value**

### BOARD INSULATION

GAFTEMT PERMALITE flat-roof insulation and tapered-roof insulation boards manufactured from perlite and recycled cellulose binders. Available in 2-ft. by 4-ft. flat panels, ¾ in. to 4 in. thick, or 2-ft. by 4-ft. tapered panels sloping 1⁄16 in., ⅛ in., 3⁄16 in., ¼ in., or ½ in. per foot. Class A fire rated. R-value is R-2.78/in. PERMALITE and RECOVER BOARD are rigid insulation boards manufactured from expanded perlite, cellulose binders (starch), and waterproofing additives. Made from 25% to 28% recycled newspaper, this product has been in use for 30 years.

## 213 CAN-CELL Industries, Inc.

14715 114 Ave., Edmonton, AB, Canada T5M 2Y8
Phone: 403 447 1255   Fax: 403 447 1034

**Price Index Number:   1.25**
**Price Index Standard:   fiberglass batt,**
                                              **overall residence**

### LOOSE-FILL INSULATION

WEATHERSHIELD is a cellulose insulation manufactured from 100% recycled newsprint, impregnated with borate to resist mold and decay and to act as a fire retardant. R-value of R-3.81/in. Settle density is 1.55 lb./cu. ft. Passed Canadian flame-spread tests. Distribution limited to the Midwest.

## 214 Cell-Pak, Inc.

PO Box 1023, Decatur, AL 35602
Phone: 800 325 5320 or 205 350 3311
Fax: 205 350 0308

**Price Index Number:   0.90**
**Price Index Standard:   fiberglass batt**

### LOOSE-FILL INSULATION

Cellulose insulation manufactured from 100% recycled newsprint with soybean-based inks. Fire-retardant formula is proprietary. Distribution limited to the Southeast.

## 215 Central Fiber Corp.

4814 Fiber Ln., Wellsville, KS 66092
Phone: 800 654 6117 or 913 883 4600
Fax: 913 883 4429

**Price Index Number:   0.90**
**Price Index Standard:   fiberglass batt**

### LOOSE-FILL INSULATION

SHELTER SHIELD CLEAN is a loose-fill insulation manufactured from 80% recycled newsprint. SUNCELL wall-cavity spray insulation is available for new construction. R-value of R-3.7/in. Settled density is 1.7 lb./cu. ft. Manufacturer claims product installs at its settled density thickness.

## 216 CertainTeed Corp. Insulation Group

PO Box 860, Valley Forge, PA 19482
Phone: 800 274 8530, 610 341 7000,
or 610 640 4104    Fax: 610 640 4130

**Price Index Number:    1.0**
**Price Index Standard:   virgin fiberglass batt**

**FIBERGLASS INSULATION**
Fiberglass batt manufactured with 20% to 25% recycled glass content, meeting EPA requirements. Made with 30% recycled content for California. Company also produces EASYHANDLER R-13, R-19, and R-30 batts wrapped in nonwoven fabric for use in walls, ceilings, and floors. R-value of R-3/in. to R-3.3/in.

## 217 Champion Insulation

PO Box 320, Lomira, WI 53048
Phone: 800 236 4310 or 414 269 4311
Fax: 414 269 4301

**Price Index Number:    0.7 material/1.0**
**                       material + labor**
**Price Index Standard:   fiberglass batt**

**LOOSE-FILL INSULATION**
Cellulose insulation made from recycled newsprint with boric acid added as a fire retardant. Distribution limited to Michigan, Wisconsin, Minnesota, Illinois, and Iowa.

## 218 Clayville Insulation

250 E. Highway 81, Burley, ID 83318
Phone: 800 584 9622 or 208 678 9791
Fax: 208 678 9784

**Price Index Number:    1.0**
**Price Index Standard:   fiberglass batt**

**LOOSE-FILL INSULATION**
Cellulose insulation made from recycled newspaper with boric acid added as a fire retardant.

## 219 Energy Pro

14235 S.E. 98th Ct., Clackamas, OR 97015
Phone: 503 653 0063    Fax: 503 653 0467

**Price Index Number:    0.90**
**Price Index Standard:   fiberglass batt**

**LOOSE-FILL INSULATION**
Cellulose insulation made from recycled newspaper with boric acid added as a fire retardant.

## 220 Energy Zone Manufacturing, Inc.

1002 1st St. N.E., Buffalo, MN 55313
Phone: 800 458 8363 or 612 682 5755
Fax: 612 682 5759

**Price Index Number:    0.90**
**Price Index Standard:   fiberglass batt**

**LOOSE-FILL INSULATION**
Cellulose insulation made from 100% recycled newsprint with boric acid added as a fire retardant. Product distribution is limited to northern Iowa, Minnesota, and eastern North Dakota.

## 221 Environmentally Safe Products, Inc.

313 W. Golden Ln., New Oxford, PA 17350
Phone: 800 289 5693 or 717 624 3581
Fax: 717 624 7089

**Price Index Number:**    **3.2**
**Price Index Standard:**   **fiberglass batt**

### FOAM INSULATION

LOW-E and MICRO-E Insulation manufactured from polyethylene foam bonded between two skins of aluminum. LOW-E is ¼ in. thick with an R-value of R-7.75 in a vertical wall cavity with air space; R-7.55 in a horizontal ceiling location with air space and upward heat flow; and R-10.74 in a horizontal ceiling location with air space and downward heat flow. MICRO-E is ⅛ in. thick, and, according to the manufacturer, has a slightly lower R-value than LOW-E but has not been tested. Available for installation in 16 in., 24 in. and 48 in. o.c. stud spacing with tabs. Product lengths are 75 ft., 100 ft., and 125 ft., respectively. Made from 40% pre- and post-consumer (plastic milk bottles) recycled foam. Approved for use in water heater jackets by the California Bureau of Home Furnishings and Thermal Insulation.

## 222 Fibrex, Inc.

PO Box 1148, Aurora, IL 60507-1148
Phone: 800 342 7391 or 708 896 4800
Fax: 708 896 3200

**Price Index Number:**    **3.0, bonded blanket**
**Price Index Standard:**   **fiberglass batt**

### MINERAL-WOOL INSULATION

Insulation blankets made from 92% recycled mineral slag, ceramic tile, and rocks. Blankets designed to fit between wall studs for both thermal and acoustical insulation. R-value of R-4/in. Contains no asbestos. It is noncombustible and noncorrosive, and it will not absorb moisture. Will not settle, rot, decay, or break down. Blankets bonded with organic phenolic resin, which will outgas at 450°F to 500°F. Also available in a loose-fill form to be blown into an attic space. Loose fill is nonbonded with a mineral oil coating, which has some combustible properties.

## 223 Foam Tech, Inc.

PO Box 8, Rt. 5, North Thetford, VT 05054
Phone: 800 374 3626 or 802 333 4333
Fax: 802 333 4364

**Price Index Number:**    **1.0-1.2**
**Price Index Standard:**   **HCFC-22 urethane foams**

### FOAM INSULATION

SUPERGREEN high-density polyurethane insulation is a field-applied foam with zero ozone depletion potential. Aged R-value is R-6.7/in. Can be used for entire house or as a spray or injection foam insulation around door and window frames and other hard-to-reach places prior to applying drywall. Uses HFC-134a containing no chlorine as a blowing agent. Manufacturer claims no CFCs or HCFCs used. It is field applied by company crew and is currently available only in northern New England. Manufactuer claims this product to be nontoxic.

## 224 GreenStone Industries, Fort Wayne

PO Box 13048, Fort Wayne, IN 46866
Phone: 800 286 8012 or 219 420 7600
Fax: 219 420 2553

**Price Index Number:**    **0.5, attic/ 1.1, wall + labor**
**Price Index Standard:**   **fiberglass batt**

### LOOSE-FILL INSULATION

R-PRO and R-PRO PLUS cellulose insulations are made from recycled newsprint with borate fire retardant and designed for various applications (loose blown in and wall spray). R-PRO PLUS has wheat-starch binder for wall applications. R-value is R-3.7/in., UL listed. Manufacturing plants are located in Arizona, California, Georgia, Indiana, Nebraska, Oregon, and Virginia.

## 225 GreenStone Industries, Maryland

6500 Rock Spring Drive, Bethesda, MD 20817
Phone: 301 564 5900　Fax: 301 564 0004

**Price Index Number:**　**0.5, attic/**
　　　　　　　　　　　　**1.1, wall + labor**
**Price Index Standard:**　**fiberglass batt**

**LOOSE-FILL INSULATION**
R-PRO and R-PRO PLUS cellulose insulations are made from recycled newsprint with borate fire retardant and designed for various applications (loose blown in and wall spray). R-PRO PLUS has wheat-starch binder for wall applications. R-value is R-3.7/in., UL listed. Manufacturing plants are located in Arizona, California, Georgia, Indiana, Nebraska, Oregon, and Virginia.

## 226 GreenStone Industries, Norfolk

PO Box 1533, East Hwy 24, Norfolk, NE 68702
Phone: 800 228 0024 or 402 379 2040
Fax: 402 379 2040

**Price Index Number:**　**0.5, attic/**
　　　　　　　　　　　　**1.1, wall + labor**
**Price Index Standard:**　**fiberglass batt**

**LOOSE-FILL INSULATION**
R-PRO and R-PRO PLUS cellulose insulations are made from recycled newsprint with borate fire retardant and designed for various applications (loose blown in and wall spray). R-PRO PLUS has wheat-starch binder for wall applications. R-value of R-3.7/in., UL listed. Manufacturing plants are located in Arizona, California, Georgia, Indiana, Nebraska, Oregon, and Virginia.

## 227 GreenStone Industries, Sacramento

5854 88th St., Sacramento, CA 95828
Phone: 800 655 9754 or 916 387 9754
Fax: 916 387 9755

**Price Index Number:**　**0.82, attic/2.43, wall**
**Price Index Standard:**　**fiberglass batt**

**LOOSE-FILL INSULATION**
R-PRO and R-PRO PLUS cellulose insulations are made from recycled newsprint with borate fire retardant and designed for various applications (loose blown in and wall spray). R-PRO PLUS has wheat-starch binder for wall applications. R-value of R-3.7/in., UL listed. Manufacturing plants are located in Arizona, California, Georgia, Indiana, Nebraska, Oregon, and Virginia.

## 228 Greenwood Cotton Insulation

PO Box 1017, Greenwood, SC 29648
Phone: 800 546 1332 or 770 998 6888
Fax: 770 998 8803

**Price Index Number:**　**1.3-1.4**
**Price Index Standard:**　**fiberglass batt**

**COTTON INSULATION**
INSUL-COT is cotton/polyester insulation. Fibers are nonirritating (chemical insect and rodent repellents and inhibitors might be a problem for the chemically sensitive). R-value of R-3.2/in. at 0.42-lb. density. No representative available on the West Coast.

## 229 Hamilton Manufacturing, Inc.

118 Market St., Twin Falls, ID 83301
Phone: 800 777 9689 or 208 733 9689
Fax: 208 733 9447

**Price Index Number:** **0.95**
**Price Index Standard:** **fiberglass batt**

### LOOSE-FILL INSULATION
Cellulose insulation made from 100% recycled waste household paper, such as mail, telephone books, and newspaper, with boric acid added as a fire retardant. Distribution is mostly in the northwestern United States.

## 230 Harborlite Corp.

PO Box 1014, LaPorte, TX 77571-1014
Phone: 800 856 3122 or 713 417 3122
Fax: 713 471 8304

**Price Index Number:** **0.50, perlite masonry fill**
**Price Index Standard:** **polystyrene beads**

### MINERAL INSULATION
Provides perlite for insulation and insulation products. Largest producer of world perlite.

## 231 Icynene, Inc.

376 Watline Ave., Mississauga, ON, Canada L4Z 1X2
Phone: 800 361 3155, 800 758 7325, or 906 890 7325   Fax: 905 890 7784

**Price Index Number:** **2.0**
**Price Index Standard:** **fiberglass batt**

### SPRAYED INSULATION
ICYNENE is a carbon dioxide foamed, modified urethane insulation. R-value of about R-3.6/in. Sprayed in open stud cavity, it sticks to everything it touches and expands to create a complete air seal. Maintains R-value without outgassing. Good acoustical insulation. In use in Canada since 1982.

## 232 In-Cide Technologies

50 N. 41st Ave., Phoenix, AZ 85009
Phone: 800 777 4569 or 602 233 0756
Fax: 602 272 4864

**Price Index Number:** **0.95**
**Price Index Standard:** **fiberglass batt**

### LOOSE-FILL INSULATION
IN-CIDE is an insecticide insulation made from recycled newsprint with a clay, starch, or acrylic latex binder and a boric acid pesticide. Available for loose-fill or spray applications. R-value is R-3.6/in. Manufacturer claims product to be the only EPA-approved pest-control insulation.

## 233 Insul-Tray

E. 1881 Crestview Dr., Shelton, WA 98584
Phone: 360 427 5930   Fax: 360 427 5930

**Price Index Number:** **1.8, west/1.3, east— panel + cellulose**
**Price Index Standard:** **fiberglass batt**

### BOARD INSULATION
INSUL-TRAY corrugated-cardboard panels are stapled into walls, ceilings, and floors to form cavities for filling with loose-fill cellulose insulation. Moisture resistant and made from 85% recycled paper. Claims R-value better than fiberglass.

## 234 Insulfoam

12601 E. 33rd Ave., #114, Aurora, CO 80011
Phone: 800 735 4621 or 303 366 7730
Fax: 303 366 7742

**Price Index Number:** 0.90, construction grade
**Price Index Standard:** virgin expanded polystyrene

### EXPANDED POLYSTYRENE
Expanded polystyrene insulation manufactured with steam without CFCs, HCFCs, or formaldehyde. R-value is R-3.85/in. Manufacturer makes a construction grade of expanded polystyrene insulation with a 15% recycled polystyrene content. UL approved. Also manufactures polystyrene void material with a 40% recycled polystyrene content for concrete formwork. Masonry block polystyrene insulation fill with a 100% recycled content is also available .

## 235 International Cellulose

12315 Robin Blvd., Houston, TX 77245-0006
Phone: 800 444 1252 or 713 433 6701
Fax: 713 433 2029

**Price Index Number:** 0.5-0.8
**Price Index Standard:** fiberglass batt

### SPRAYED INSULATION
CELBAR is a residential insulation made from recycled newsprint and is available as a wall spray with a water based adhesive or a loose fill for attics. Borate is added as a fire retardant. R-value is R-3.8/in. UL listed. K13 and K13FC are commercial insulations for rough and finished ceilings, respectively. Available in various colors including custom tints. K13FC maximum thickness is 1 in., with an R-value of R-4.54. K13 maximum thickness is 4 in. Borax added as a fire retardant. UL listed.

## 236 Louisiana-Pacific

7825 Trade St., #100, San Diego, CA 92121
Phone: 800 917 2077    Fax: 619 586 0337

**Price Index Number:** 0.90
**Price Index Standard:** fiberglass batt

### LOOSE-FILL INSULATION
NATURE GUARD insulation is made from 100% recycled newsprint. R-value is R-3.8/in. Settle density is 1.6 lb./cu. ft. Manufacturer claims product contains longer fibers than most recycled newsprint products, making it more resilient and more efficient.

## 237 Modern Insulation

1206 S. Monroe St., Spencer, WI 54479
Phone: 800 945 0186 or 715 659 2446
Fax: 715 659 4734

**Price Index Number:** 0.60-0.75
**Price Index Standard:** blown-in fiberglass

### LOOSE-FILL INSULATION
Cellulose insulation from 100% recycled newsprint with boric acid added as a fire retardant. Distribution limited to the Midwest.

## 238 Mountain Fiber Insulation

PO Box 337, Hyrum, UT 84319-0337
Phone: 800 669 4951 or 801 245 6081
Fax: 801 245 4475

**Price Index Number:** 0.74, material/1.0, material + labor
**Price Index Standard:** fiberglass batt

### LOOSE-FILL INSULATION
Cellulose insulation for attics made from recycled newsprint with boron added as a fire retardant. Made from 80% recycled paper and 20% boron by weight. Cellulose wall insulation uses sulfate-based paper for a higher R-value and wet-spray applications. R-value is 3.7/in.

## 239 Northern Insulation Products

414 E. 13th St., PO Box 26, Gibbon, MN 55335
Phone: 800 722 0543 or 507 834 6519
Fax: 507 834 6518

**Price Index Number:** **0.63, material/1.30, material + labor**
**Price Index Standard:** **fiberglass batt**

**LOOSE-FILL INSULATION**
Cellulose insulation made from recycled newsprint with boric acid and borax. Distribution limited to Minnesota, North Dakota, South Dakota, Iowa, and Wisconsin.

## 240 Nu-Woll Insulation

2472 Port Sheldon Rd., Genison, MI 49428
Phone: 800 748 0128 or 616 669 0100
Fax: 616 669 2370

**Price Index Number:** **0.63, material/1.30, material + labor**
**Price Index Standard:** **fiberglass batt**

**LOOSE-FILL INSULATION**
Cellulose insulation made from recycled newsprint with boric acid added as a fire retardant. Distribution limited to Wisconsin, Illinois, Ohio, Kentucky, and Michigan.

## 241 Ottawa Fibre, Inc.

3985 Balgreen, Ottawa, ON, Canada K1G 3N2
Phone: 613 736 1215
Fax: 613 736 1150 (info) or 613 731 3865 (order)

**Price Index Number:** **1.0**
**Price Index Standard:** **virgin fiberglass batt**

**FIBERGLASS INSULATION**
Fiberglass insulation products made from 100% recycled glass with an overall recycled content of 60% to 80%. Used in walls, ceilings, and floors. R-value is R-3.3/in. to R-4/in.

## 242 Owens Corning

One Fiberglass Tower, Toledo, OH 43659
Phone: 800 766 3464, 800 235 4599, or
419 248 8000

**Price Index Number:** **0.95**
**Price Index Standard:** **fiberglass batt insulation, R-19**

**FIBERGLASS INSULATION**
MIRAFLEX R-25 fiberglass insulation is made by fusing two types of glass fibers together. The fibers twist about each other to creat a mechanical bond, eliminating the need for binders. This results in a more resilient and stronger fiber with less of a tendency to get into the air or to irritate skin. MIRAFLEX also packs better for shipping, requiring only one-quarter the volume of comparable standard fiberglass batt, then expanding to 8¾ in. when installed.

## 243 P. K. Insulation Manufacturing Co.

PO Box 281, Joplin, MO 64802
Phone: 800 641 4296   Fax: 417 781 8335

**Price Index Number:** **0.63, material/1.30, material + labor**
**Price Index Standard:** **fiberglass batt**

**LOOSE-FILL INSULATION**
Cellulose insulation made from 100% recycled newsprint with boric acid added as a fire retardant. Material is applied with wet spray. Distribution is mostly in the Midwest.

## 244 Palmer Industries

10611 Old Annapolis Rd., Frederick, MD 21701
Phone: 301 898 7848   Fax: 301 898 3312

**Price Index Number:**   1.9-3.8
**Price Index Standard:**   fiberglass batt

### FOAM INSULATION

AIR-KRETE is a cementitious thermal and acoustical foam insulation designed for hose applications. Applied to new or existing structures with a hose into an enclosed or open cavity. Open-cavity application, with screen, is troweled in place. AIR-KRETE does not expand after leaving hose. Quality not affected by installation temperature. Manufacturer claims a 3½-in. thickness of AIR-KRETE has a 3-hour fire rating. At a typical installation density of 2 lbs./cu. ft., the R-value is R-3.9/in. The product is a concrete-like material containing seawater-derived magnesium oxide. N.Y. Housing Authority is currently installing AIR-KRETE in its housing projects for purposes of fireproofing. Product is 98% inorganic, chemically stable with no outgassing, and has less than ¼% of 1% shrinkage. Manufacturer claims product to be nontoxic and maintenance free.

## 245 Paul's Insulation

PO Box 115, Vergas, MN 56587
Phone: 800 627 5190   Fax: 218 342 3050

**Price Index Number:**   0.90
**Price Index Standard:**   fiberglass batt

### LOOSE-FILL INSULATION

Cellulose insulation made from recycled newsprint with boric acid added as a fire retardant. Available only in dry-pack form. Distribution limited to North Dakota, South Dakota, and Minnesota.

## 246 Perma Flake Corp.

PO Box 4653, Greenville, MS 38704
Phone: 601 334 9852

**Price Index Number:**   0.63, material/1.30, material + labor
**Price Index Standard:**   fiberglass batt

### LOOSE-FILL INSULATION

Cellulose insulation made from recycled cardboard-cosmetic boxes with boric acid added as a fire retardant. Dry-pack or wet spray-on application. Distribution limited to Mississippi.

## 247 Persolite Products, Inc.

PO Box 505, Florence, CO 81226
Phone: 800 872 1973, 719 784 6531
or 303 572 3222   Fax: 719 784 4855

**Price Index Number:**   0.50, perlite masonry fill
**Price Index Standard:**   polystyrene beads

### MINERAL INSULATION

Perlite volcanic-rock insulating material in the form of small balls for insulating concrete block. Typical 8-in. concrete block with perlite insulation has an R-value of R-8.25. Noncombustible.

## 248 Redco II

11831 Vose, N. Hollywood, CA 91605
Phone: 818 759 2255   Fax: 818 503 1319

**Price Index Number:** **0.40, perlite masonry fill**
**Price Index Standard:** **polystyrene beads**

**MINERAL INSULATION**
Provides perlite for insulation and insulation products. Available bagged or in bulk.

## 249 Regal Industries, Inc.

9564 E. Country Rd. 600 S., Crothersville, IN 47229
Phone: 800 848 9687 or 812 793 2214
Fax: 812 793 3432

**Price Index Number:** **0.90**
**Price Index Standard:** **fiberglass batt**

**LOOSE-FILL INSULATION**
Cellulose made from 100% recycled newsprint. Fire retardant-treated with borates but should not be applied where temperatures exceed 180°F. Blown into attic and wall cavities for best R-value. R-value is R-2.92/in. Settle density is 1.7 lb./in. Noncorrosive, nontoxic. UL listed.

## 250 Schuller International, Inc.

PO Box 5108, Denver, CO 80217-5108
Phone: 800 654 3103 or 303 978 2785
Fax: 303 978 3661

**Price Index Number:** **1.0/1.20, with polyethylene wrap**
**Price Index Standard:** **virgin fiberglass batt**

**FIBERGLASS INSULATION**
Manufacturer of fiberglass insulation, including batts, blankets, board, and loose fill. Claims 20% to 30% recycled glass content. R-value is R-3/in. to R-3.3/in. Also produces CONFORTTHERM R-11, R-13, R-15, R-19, and R-25 fiberglass batts wrapped in polyethylene for health reasons.

## 251 Schuller International, Inc.

PO Box 5108, Denver, CO 80217-5108
Phone: 800 654 3103   Fax: 303 978 3904

**Price Index Number:** **2.0**
**Price Index Standard:** **polyiso per R-value**

**BOARD INSULATION**
FESCO BOARD roofing system manufactured from perlite and cellulose fibers from 25% to 35% recycled paper. R-value is R-2.78/in.

## 252 Suncoast Insulation Manufacturing Co.

7102 N. 30th St., Tampa, FL 33610
Phone: 800 666 4824 or 813 238 0486
Fax: 813 234 9741

**Price Index Number:** **0.63, material/1.30, material + labor**
**Price Index Standard:** **fiberglass batt**

**LOOSE-FILL INSULATION**
Cellulose insulation made from recycled newsprint and telephone books with boric acid added as a fire retardant.

## 253 Tascon, Inc.

PO Box 41846, Houston, TX 77241
Phone: 800 937 1774 or 713 937 0900
Fax: 713 937 1496

**Price Index Number:** **0.6, material/1.30, material + labor**
**Price Index Standard:** **fiberglass batt**

### LOOSE-FILL INSULATION
Cellulose loose-fill insulation with boric acid added as a fire retardant. R-value of R-3.7/in. or R-13.25 for 4 in. Made from 100% recycled newsprint. Distribution limited to Texas, Oklahoma, Louisiana, Arkansas, Alabama, Mississippi, and western Florida.

## 254 Tenneco Building Products

2907 Log Cabinet Dr., Smyrna, GA 30080
Phone: 800 241 4402    Fax: 404 350 1489

**Price Index Number:** **1.4, per R-value comparison**
**Price Index Standard:** **fiberglass batt**

### BOARD INSULATION
AMOFOAM-RCY extruded polystyrene insulation board from at least 50% recycled material. Foamed with HCFCs rather than CFCs to minimize ozone depletion in the upper atmosphere. Used as foundation edge, cavity wall, and exterior insulation. R-value is R-5/in. Available in 1-in., 1½-in., and 2-in. thicknesses in 8-ft. by 2-ft. or 4-ft. sheets.

## 255 Tennessee Cellulose, Inc.

60 Davy Crockett Park Rd., Limestone, TN 37681
Phone: 423 257 2051    Fax: 423 257 4821

**Price Index Number:** **0.5, material/1.40, material + labor**
**Price Index Standard:** **fiberglass batt**

### LOOSE-FILL INSULATION
Cellulose loose fill insulation with boric acid added as a fire retardant. R-value is R-3.7/in. and R-13.25 for 4 in. Made from 100% recycled newsprint. Distribution limited to Tennessee, Kentucky, Virginia, North Carolina, South Carolina, and West Virginia.

## 256 ThermaFiber LLC

2301 Taylor Way, Tacoma, WA 98421
Phone: 800 426 8127 or 206 627 0379
Fax: 206 627 0424

**Price Index Number:** **0.50**
**Price Index Standard:** **fiberglass batt**

### MINERAL-WOOL INSULATION
Loose-fill insulation of mineral fibers (rock wool) with 50% to 70% recycled content for residential and commercial use. R-value is R-3/in. to R-3.4/in. Contains no asbestos. It is noncombustible, noncorrosive, and will not absorb moisture. Will not settle, rot, decay, or break down. Vermin proof.

### 257 Thermoguard Insulation Co.

451 Charles St., Billings, MT 59101
Phone: 800 821 5310 or 406 252 1938
Fax: 406 252 5019

**Price Index Number:   0.90**
**Price Index Standard:   fiberglass batt**

**LOOSE-FILL INSULATION**
ISOLITE is a blow-in cellulose insulation made from 100% recycled newsprint with boric acid added as a fire retardant. R-value is R-3.6/in. Settle density is 2.0 to 2.3 lbs./cu. ft.

### 258 Universal Polymer

319 N. Main Ave., Springfield, MO 65806
Phone: 800 752 5403   Fax: 417 862 7548

**Price Index Number:   0.60/0.80**
**Price Index Standard:   Dryvit/3-coat stucco**

**BOARD INSULATION**
EXCEL BOARD is an aspen wood-fiber mat encapsulated in an isocyanurate foam, without VOCs. Considerably stronger than foam sheathing with a 5-lb. density and kraft paper (with aluminum foil) backing, though not as hard as plywood or particleboard. Intended to be a very competitive substrate for EIF systems and low-cost synthetic stucco systems. Although the specs should be quite good, fire rating and insulation value has not been tested yet.

### 259 West Materials, Inc.

PO Box 1531,
Burnsville, MN 55337-1531
Phone: 612 892 7305   Fax: 612 892 7573

**Price Index Number:   1.7**
**Price Index Standard:   concrete block with perlite insulating cores**

**POLYSTYRENE INSERTS**
ENERBLOCK molded expanded polystyrene insulation inserts for the voids within and between standard concrete masonry units. Available for block sizes 8 in., 10 in., or 12 in. wide, and insert widths of 1¼ in., 2 in., 2½ in., or 4 in. R-value varies from R-3 to R-10.2 depending upon block and insert width combinations. Inserts are designed to accommodate reinforcing steel. Fire rating is flame spread 15 and smoke developed 45 to 125. Available in the upper Midwest.

## ROOF AND DECK INSULATION                               07220

### 260 Huebert Fiberboard Co.

E. Morgan St., PO Box 167, Boonville, MO 65233
Phone: 800 748 7147 or 816 882 2704
Fax: 816 882 7991

**Price Index Number:   2.0**
**Price Index Standard:   polyiso per R-value**

**BOARD INSULATION**
Two types of wood-fiber roof insulation RECOVER BOARD manufactured from recycled magazines and newspapers (15%) and waste mill wood chips (85%). For use on flat roofs with single-ply membrane systems requiring a Class E underlayment. ROOF INSULATION BOARD is available in 4-ft. by 2-ft., 4-ft., and 8-ft. sheets in ½-in., ¾-in., and 1-in. solid thicknesses, and 1½-in. and 2-in. laminated thicknesses. HIGH DENSITY ROOF INSULATION PLUS is available in 2-ft. by 4-ft. by ½-in. thick sheets with a baked on water-based asphalt emulsion coating. R-value is R-2.78/in.

### 261 Persolite Products, Inc.

PO Box 505, Florence, CO 81226
Phone: 719 784 6531 or 303 572 3222
Fax: 719 784 4855

**Price Index Number:   1.50**
**Price Index Standard:   concrete**

**MINERAL INSULATION**
Manufacturer of lightweight, insulating concrete aggregate made from perlite for all building applications.

## 262 American Sprayed Fibers, Inc.

1550 E. 91st Dr., Merrillville, IN 46410
Phone: 800 824 2997 or 219 769 0180
Fax: 219 736 6126

**Price Index Number:** **0.3**
**Price Index Standard:** **cementitious**
**fireproofing**

### SPRAY FIREPROOFING
DENDAMIX is made from a mixture of recycled paper, stone, and plastic.

## 263 American Cemwood Products

PO Box C, Albany, OR 97321
Phone: 800 367 3471 or 541 928 6397
Fax: 541 928 8110

**Price Index Number:** **6.5**
**Price Index Standard:** **composition shingle,**
**20 year**

### FIBER-CEMENT SHINGLES
CEMWOOD and PERMATEK shakes are manufactured with two-thirds Portland cement and one-third wood fiber from waste sawmill scraps. These are lightweight, noncombustible, composite products that can be worked with hand tools and nailed. CEMWOOD shakes weigh 580 lbs./square, and PERMATEK, 450 lbs./square. Come with a 50-year warranty.

## 264 ATAS International, Inc.

6612 Snowdrift Dr., Allentown, PA 18106
Phone: 800 468 1441   Fax: 610 395 9342

**Price Index Number:**   **0.60, 29-gauge steel**
**Price Index Standard:**   **slate roofing tiles**

### METAL SHINGLES
Steel or aluminum shingles with more than 60% recycled content. Most such metal products are made from industrial remelt. Available in 10 colors with an option for custom colors. Weight is 35 lbs./square to 60 lbs./square for aluminum and 80 lbs./square to 140 lbs./square for steel. Fire rating can be made to be Class A with appropriate substrate or underlayment. Comes with a 25-year warranty.

## 265 ATAS International, Inc.

4559 Federal Blvd., San Diego, CA 92102
Phone: 800 879 8382 or 619 696 0102
Fax: 619 262 9914

**Price Index Number:**   **0.60, 29-gauge steel**
**Price Index Standard:**   **slate roofing tiles**

### METAL SHINGLES
Steel or aluminum shingles with more than 60% recycled content. Most such metal products are made from industrial remelt. Available in 10 colors with an option for custom colors. Weight is 35 lbs./square to 60 lbs./square for aluminum, and 80 lbs./square to 140 lbs./square for steel. Fire rating can be made to be Class A with appropriate substrate or underlayment. Comes with a 25-year warranty.

## 266 Atlas Roofing Corp.

PO Box 5777, Valley Rd., Meridian, MS 39302
Phone: 800 933 2721 or 601 484 8900
Fax: 601 483 7844, 405 223 6894,
513 746 1528, or 215 536 3002

**Price Index Number:**   **1.20**
**Price Index Standard:**   **fiberglass shingles**

### ROOFING SHINGLES
Organic roofing shingles manufactured from 100% recycled paper and asphalt, and finished with a durable asphalt coating covered with stone granules. Class C fire rating. Product performs best in warmer climates.

## 267 California Shake Corp.

5355 N. Vincent Ave., Irwindale, CA 91706
Phone: 818 812 9085 or 818 969 7544
Fax: 818 969 7124

**Price Index Number:** **7.7, shake/8.2, slate**
**Price Index Standard:** **composition shingle,**
**20 year**

### FIBER-CEMENT SHINGLES

Fiber-cement tiles with perlite made in shake, clay and slate styles and natural colors, measuring 22 in. by 12 in. by ⅞ in. R-value is R-2.4. Class A fire rated. Weight is 5.6 lbs./sq. ft. Comes with 25-year limited, transferable product warranty. Not recommended for use in areas with high freeze/thaw cycles.

## 268 Classic Products, Inc.

299 Staunton St., PO Box 701, Piqua, OH 45356
Phone: 800 543 8938 or 513 773 9840
Fax: 513 773 9261

**Price Index Number:** **7.3**
**Price Index Standard:** **composition shingle,**
**20 year**

### METAL SHINGLES

RUSTIC SHINGLE is an aluminum shingle, 12 in. by 24 in., made from recycled aluminum beverage cans. Textured and colored to look like wood shakes. Manufacturer claims product is made from 98% post-consumer material. This product weighs 45 lbs./square. Comes with a 20-year/50-year warranty. UL Class A.

## 269 Crowe Industries Ltd.

116 Burris St., Hamilton, ON, Canada L8M 2J5
Phone: 905 529 6818   Fax: 905 529 1755

**Price Index Number:** **0.80-1.3**
**Price Index Standard:** **slate roofing tiles**

### POLYMER SHINGLES

AUTHENTIC ROOF slates, manufactured to imitate black slate, are made from 100% recycled polymers and rubber. This product weighs 50 lbs. to 80 lbs. less per square than asphalt shingles. Product is completely recyclable and comes with a 50-year warranty.

## 270 Eternit

PO Box 679, Blandon, PA 19510-0679
Phone: 800 233 3155 or 610 926 0100
Fax: 610 926 9232

**Price Index Number:** **0.80**
**Price Index Standard:** **slate roofing tiles**

### FIBER-CEMENT SHINGLES

Slate finish, predrilled fiber-cement tiles. Available in 2 sizes and 5 colors.

## 271 GAF Corp.

1361 Alps Rd., Wayne, NJ 07470
Phone: 800 223 1948   Fax: 201 628 3865

**Price Index Number:** **1.5**
**Price Index Standard:** **wood shakes, installed**

### FIBER-CEMENT SHINGLES

Portland cement and cellulose reinforced shingle available in slate, shake, and traditional styles. Available in rectangular, square, hexagonal, and pentagonal shapes, 9.35 in. by 16 in., in a variety of colors. Weight is from 240 lbs./square to 500 lbs./square, depending on style. Fire rating is Class A or B. Predrilled nail holes. Comes with a 40-year nonprorated, limited warranty.

## 272 Georgia-Pacific Corp.

133 Peachtree St. N.E., Atlanta, GA 30303
Phone: 800 284 5347 or 404 652 4000
Fax: 800 830 7385 (West) or 800 839 2595 (East)

**Price Index Number:** **1.25**
**Price Index Standard:** **fiberglass**
 **composition shingles**

### ORGANIC SHINGLES

Organic asphalt shingles with recycled paper in many colors. SUMMIT III is an architectural series 40-year roof, weighing 300 lbs./square and having a Class A fire rating. SUMMIT is the same thickness as a SUMMIT III with less density for a 30-year roof weighing 250 lbs./square and having a Class A fire rating. T-LOCK is a 20-year roof that weighs 245 lbs./square and has a Class C fire rating.

## 273 Gerard Roofing Technology

955 Columbia St., Brea, CA 92821-2923
Phone: 800 841 3213 or 714 529 0407
Fax: 714 529 6643

**Price Index Number:** **0.40**
**Price Index Standard:** **slate roofing tiles**

### METAL SHINGLES

Roofing tiles and shakes made of prepainted galvanized or galvalume (zinc/aluminum) 26-gauge steel, coated with graded stone granules. Most such metal products are made from industrial remelt. Weigh 140 lbs./square. Fire rating Class A. Interlocking panels create a continuous weather barrier to withstand 120 mph winds. Comes with a 50-year limited warranty.

## 274 Globe Building Materials, Inc.

2230 Indianapolis Blvd., Whiting, IN 46394
Phone: 800 950 4562 or 219 473 4500
Fax: 219 473 3504

**Price Index Number:** **1.25**
**Price Index Standard:** **fiberglass shingle**

### ROOFING SHINGLES

GLOBE SUPER SEAL organic roofing shingles manufactured from 68% recycled paper, and asphalt, and finished with a durable coating. Class C fire rating. Performs best in warmer climates.

## 275 James Hardie Building Products

10901 Elm Ave., Fontana, CA 92335
Phone: 800 942 7343 or 909 356 6300
Fax: 909 355 0690

**Price Index Number:** **6.8, HARDISHAKE/**
 **8.6, HARDISLATE**
**Price Index Standard:** **composition**
 **shingles, 20 year**

### FIBER-CEMENT SHINGLES

HARDISHAKE and HARDISLATE fiber-cement composite roofing slates are manufactured from Portland cement and wood-fiber cellulose. Comes in 7 colors and has a weight of 380 lbs./square. Can achieve a Class A fire rating. Has a 50-year warranty in the sunbelt states, such as California, Arizona, and Florida.

## 276 Masonite Corp.

1 South Wacker Dr., Chicago, IL 60606
Phone: 800 323 4591 or 800 647 7080
Fax: 630 584 6349

**Price Index Number:** **0.60**
**Price Index Standard:** **No. 1 cedar shingles**

### WOOD SHINGLES

WOODRUF traditional roofing shingles from heat-bonded wood fibers. More dense and durable than natural wood. Available in natural brown wood tones to look like cedar. Size is 1 ft. by 4 ft. by 7/16 in., similar to a medium blue cedar shingle. Weight is 230 lbs./square. No field finishing required and can be installed with standard carpenter's tools. Comes with a 15-year limited warranty. Does not meet minimum Class C fire rating for roofing applications in California.

## 277 MaxiTile, Inc.

17141 S. Kingsview Ave., Carson, CA 90746
Phone: 800 338 8453   Fax: 310 515 6851

**Price Index Number:** 5.4
**Price Index Standard:** composition
shingles, 20 year

### FIBER-CEMENT SHINGLES

MAXITILE is manufactured from Portland cement, silica, and cellulose fiber into a composite roofing tile to look like a 2-piece Mission tile. Available in 36½-in. by 24-in. shingle with hip and ridge caps, ¼ in. thick. Weight is 340 lbs./square. Fire rating is Class A, with no asbestos. Available in 5 earth-tone colors. Comes with a transferable 50-year limited product warranty.

## 278 Metal Sales Manufacturing Corp.

6260 Downing, Denver, CO 80216
Phone: 800 289 7663   Fax: 303 289 1605

**Price Index Number:** 1.0
**Price Index Standard:** split-wood shake

### METAL SHINGLES

STILE metal roofing panels are manufactured from recycled metal to look like clay roofing tiles. Available in 4-ft. widths by 3-ft. to 16-ft. lengths in 1-ft. increments, and in 5 colors. Weight is 100 lbs./square. Fire rating is Class A. Panels are hot-dipped galvanized steel covered with an epoxy-based undercoat and a primed and painted top surface. Comes with a 20-year limited warranty.

## 279 Tamko Roofing Products, Inc.

PO Box 1404, Joplin, MO 64802-1404
Phone: 800 641 4691 or 417 624 6644
Fax: 417 624 8935

**Price Index Number:** 1.0
**Price Index Standard:** composition
shingles, 20 year

### ROOFING SHINGLES

Roofing shingles manufactured from recycled paper, recycled wood pulp, and asphalt, and finished with a durable asphalt coating. Available in 3-tab and laminated shingles, product specification number ASTM-D 225 Type III or ASTM-D 3161 Type I.

## 280 Zappone Manufacturing

N. 2928 Pittsburg, Spokane, WA 99207
Phone: 800 285 2677 or 509 483 6408
Fax: 509 483 8050

**Price Index Number:** 1.0/1.5
**Price Index Standard:** slate roof/40-year
composition shingles

### METAL SHINGLES

Aluminum and copper shingles manufactured from recycled metal. Interlocking design permits installation on a 3/12 pitch roof. Available in one dimension, 15 in. by 9⅛ in. Aluminum shingle is made from 100% recycled metal, comes in 7 colors with oven-baked paint finish, is wood-grain embossed, and weighs 42 lbs./square. Aluminum shingle has a 50-year guarantee with a 20-year warranty for the finish. Copper shingle is manufactured from at least 85% recycled metal and weighs 94 lbs./square. Both products are fire rated Class A, B, and C. Concealed nailing detail with flange tested to seal against wind-driven rain to 110 mph.

## MANUFACTURED ROOFING AND SIDING 07400

## 281 AFM Corp.

24000 W. Highway 7, #201,
Excelsior, MN 55331-0246
Phone: 800 255 0176 or 612 474 0809
Fax: 612 474 2074

**Price Index Number:** 1.0+
**Price Index Standard:** stick framing with
batts

### MANUFACTURED WALL PANELS

R-CONTROL wall or roof panel constructed with EPS foam-core between two OSB facings. 5 standard sizes provide R-values from R-14.88 to R-44.71 at 75°F. Panels are stable and contain no CFCs, HCFCs, or formaldehyde. Available in sizes from 8 ft. by 4 ft. to 28 ft. with thicknesses of 3½ in., 5½ in., 7½ in., or 11½ in. Factory-cut electrical chases. Has a 20-year warranty on R-value of insulating core. This product will not twist, rack, or warp like wood and has been in use for 30 years.

## 282 Agriboard Industries

1500 S. Main St., Fairfield, IA 52556
Phone: 515 472 0363   Fax: 515 472 0018

**Price Index Number:** **0.90**
**Price Index Standard:** **residential stick framing**

### MANUFACTURED WALL PANELS

Stress-skin panels designed as an insulated structural load-bearing wall, floor, and roof system. Sandwich panel of $\frac{7}{16}$-in. OSB with a core of 100% compressed wheat straw. No glues. Comes in 8⅛-in. by 8-ft. by 16-ft. panels with an R-value of 20. 2-hour fire rating. Will span 16 ft. at 40-lbs./sq. ft. live load. Requires a crane for installation. Channels are provided for wiring, and windows are precut at the factory. Manufacturer claims product to be nontoxic.

## 283 Bellcomb Technologies

70 N. 22nd Ave., Minneapolis, MN 55411
Phone: 612 521 2425   Fax: 612 521 2376

**Price Index Number:** **1.0+**
**Price Index Standard:** **stick framing with batts**

### MANUFACTURED WALL PANELS

Interlocking phenolic-impregnated, kraft-paper honeycomb panels that can be faced with different materials. Core is 5% paper and 95% air. Structurally able to cover large spans, the system reduces lumber required for construction of a typical home by 67%. R-value equivalency of 14 for a 5½-in. panel. For floors, walls, and roofs. Can span up to 16 ft. with loads of 40 lbs./sq. ft. Precut at the factory, and assembled with glue and screw guns on-site. No formaldehyde.

## 284 Eagle Panel Systems

PO Box 748, Florissant, MO 63032
Phone: 800 643 3786 or 314 653 0205
Fax: 800 643 3786

**Price Index Number:** **0.75**
**Price Index Standard:** **2x4 framed wall with insulation**

### MANUFACTURED WALL PANELS

Insulated foam-core panels for floor, wall, or roof. Expanded polystyrene between two ⅜-in. OSB skins to form 4-ft. by 8-ft. to 24-ft. by 4-in., 6-in., and 8-in.-thick panels. Panels assembled on-site with splines for excellent reduced infiltration and high R-values of 17 to 33 depending upon thickness. Wiring chases are precut. Technology adapted from commercial refrigeration industry.

## 285 Enercept, Inc.

3100 9th Ave. S.E., Watertown, SD 57201
Phone: 800 658 3303 or 605 882 2222
Fax: 605 882 2753

**Price Index Number:** **1.02-1.05**
**Price Index Standard:** **stick framing with batts**

### MANUFACTURED WALL PANELS

ENERCEPT is a super-insulated building system that includes basement, wall, and roof panels plus beams in one package. Panels are expanded polystyrene laminated to various sheathing materials. Wall panels have 5⅝-in. EPS core insulation, and roof panels have 7⅜ in. EPS core insulation. Wiring chases are predesigned and provided by manufacturer. System is stronger than standard 2x6 framing and is comparable to conventional house construction with R-40 insulation.

## 286 Eternit

PO Box 679, Blandon, PA 19510-0679
Phone: 800 233 3155 or 610 926 0100
Fax: 610 926 9232

**Price Index Number:** **1.0/8.0**
**Price Index Standard:** **cedar siding/gypsum wall board**

### FIBER-CEMENT PANELS

Fiber-cement composite panels. ETERNIT SIDING is a fiber-cement lap siding textured to look like wood. It's nailable, paintable and stains evenly to resemble wood. ETERSPAN is an autoclaved fiber-cement building panel for damp locations. PROMAT is a calcium silicate panel used for fireproofing. Products are used for fascias, soffits, facades, skins for laminated panels, and interior and exterior walls. Available in standard 4-ft. by 8-ft. and 10-ft. sheets in many thicknesses. Products can be worked with traditional hand power tools and are dimensionally stable when exposed to moisture and temperature variations. Completely nonflammable.

## 287  Tenneco Packaging/Hexacomb

9700 Bell Ranch Dr.,
Santa Fe Springs, CA 90670-2981
Phone: 800 323 9163   Fax: 213 802 8924

**Price Index Number:** **2.2, honeycomb panel with kraft paper facing**

**Price Index Standard:** **2x4 stud partition wall**

### HONEYCOMB CORES

Hexacomb has been in business for 30 years and has provided both construction and assembly equipment for panelized housing manufacturers. Currently the only manufacturer of structural kraft-paper, honeycomb core material for laminated structural panels. Used for nonstructural floor, roof and wall panels. Greater strength and resistance to moisture, decay, rot, fungi, termites, and insects achieved with phenolic resin-impregnated honeycomb cells. Panels available in thicknesses up to 4 in. by 60 in. wide in unlimited lengths. Also available with regular kraft paper facings and vertical fluted cell structure panels. Varying amounts of recycled paper content in product, but client can custom order 100% recycled content.

## 288  Futurebilt, Inc.

A-104 Plaza del Sol, Wimberley, TX 78676
Phone: 800 487 5722 or 512 847 5721
Fax: 512 847 8845

**Price Index Number:** **1.0**

**Price Index Standard:** **stick framing with batts**

### MANUFACTURED WALL PANELS

Expanded polystyrene core between two structural wafer-board stress skins. Application for floors, walls, and roofs. Comes in 8-ft. by 4-ft. to 28-ft. long panels in thicknesses of 4½ in., 6½ in., 8¼ in., 10¼ in., and 11¼ in. R-value is R-15 to R-45, depending upon thickness. Wiring chases provided by manufacturer.

## 289  Georgia-Pacific Corp.

133 Peachtree St. N.E., Atlanta, GA 30303
Phone: 800 284 5347 or 404 652 4000
Fax: 800 830 7385 (West) or 800 839 2595 (East)

**Price Index Number:** **1.5**

**Price Index Standard:** **select heart redwood**

### FIBER-CEMENT PANELS

PRIMETRIM exterior and interior trim manufactured from a high resin, high temperature-cured, all-wood fiber composite. Used for nonstructural applications, such as fascia, rake board, corner board, band board, and trim. Comes in 16-ft. lengths without joints in 8 widths from 3½ in. to 11¼ in. and 2 thicknesses, ⅝ in. and 1 in. Factory-primed face and 2 edges. Dimensionally stable and resists warping, cupping, twisting, splitting, and checking. No isocyanurate or urea formaldehyde resins present in product. Can be worked with power hand tools.

## 290  Harmony Exchange

2700 Big Hill Rd., Boone, NC 28607
Phone: 800 968 9663 or 704 264 2314
Fax: 704 264 4770

**Price Index Number:** **1.20, material only**

**Price Index Standard:** **stick framing and batts**

### MANUFACTURED WALL PANELS

PERMA-R, INSULSPAN, and APACHE are manufacturer's representatives for insulated foam-core panels for floor, wall, or roof. EPS between two OSB or gypsum-board skins to form 4-ft. by 8-ft., 4-ft. by 24-ft., or 8-ft. by 8-ft. panels with thicknesses of 3⅜ in. to 9⅜ in. Panel R-values of 18 to 40, depending upon thickness. Manufacturing facilities are located in Tennessee, South Carolina, Michigan, and Colorado.

## 291 Homasote Co.

PO Box 7240, West Trenton, NJ 08628-0240
Phone: 800 257 9491 or 609 883 3300
Fax: 609 530 1584

**Price Index Number:**   **0.95**
**Price Index Standard:**   **2x8 framed roof with insulation**

### MANUFACTURED WALL PANELS
TUPS stress-skin panels are designed as an insulated structural load-bearing roof system. Consists of a panel of ½-in. Homasote nail base, board-top surface and a 440 Homasote board-bottom surface, with a core of isocyanurate foam. Top and bottom skins are made of 100% recycled newsprint cellulose. Come in 4-ft. by 8-ft., 10-ft., and 12-ft. panels. R-value is R-20.6 for 4-in. panel and R-29 for 5 in. Also available with bottom surface UL Class A rated for fire protection. Tongue and groove along long edge and less than 2% dimensional variation.

## 292 J-Deck, Inc. Building Systems

2587 Harrison Rd., Columbus, OH 13204
Phone: 614 274 7755   Fax: 614 274 7797

**Price Index Number:**   **1.0**
**Price Index Standard:**   **2x4 framed wall with insulation**

### MANUFACTURED WALL PANELS
EPS core between two structural wafer board stress skins. Installed using standard 2-in. by 4-in. studs at 4 ft. o.c. for structural panel system. For roof decks and exterior walls. Comes in 4-ft. by 8-ft., 9-ft., and 10-ft. sizes with thicknesses of 3⅝ in., 5⅝ in., 7⅜ in., and 9⅜ in. Average R-value of R-4/in. For roof decks and exterior walls. Windows installed by cutting out openings with a saw. Dimensionally stable and guaranteed not to delaminate.

## 293 James Hardie Building Products

10901 Elm Ave., Fontana, CA 92335
Phone: 800 942 7343 or 909 356 6300
Fax: 909 355 4907

**Price Index Number:**   **0.3, HARDIPLANK**
**Price Index Standard:**   **redwood lap siding**

### FIBER-CEMENT PANELS
Autoclaved fiber-cement composite panels manufactured from Portland cement, ground sand, and cellulose fiber. HARDITEX SHEETS, HARDIPANEL, and HARDISOFFIT are alternatives to stucco. HARDIPLANK lap siding is 6 in. to 12 in. by 12 ft. Products are worked with standard hand tools and attached with screws or nails. Waterproof, noncombustible with Class A fire rating, the products include post-consumer recycled newsprint. Have a 50-year warranty.

## 294 Lincoln Environmental Barrier System

PO Box 29, Canastota, NY 13032
Phone: 315 697 7224

**Price Index Number:**   **1.0+**
**Price Index Standard:**   **conventional framed residence**

### MANUFACTURED WALL PANELS
BARRIER SYSTEM replaces conventional framing and reduces labor costs. All components are supplied, including rigid, foil-faced polyisocyanurate insulation panels assembled with I-studs at the site to form a wall system. The system has 2 air spaces within each stud space for both acoustical and thermal insulation. The I-stud design uses less wood than conventional framing. The panels are made from 40% recycled plastic and 100% recycled aluminum. Applicable to exterior walls, the panels have an R-value of R-33.8. I-studs are 7¼ in. wide. This wall system includes framing, insulation, vapor barrier, and building wrap as an integrated unit.

## 295 Louisiana-Pacific

PO Box 1525, Lake Oswego, OR 97035
Phone: 800 579 8401 (order), 800 777 0749 or 503 221 0800   Fax: 503 624 9044

**Price Index Number:**   **0.4**
**Price Index Standard:**   **redwood lap siding**

### HARDBOARD SIDING
INNER-SEAL lap and panel siding consisting of oriented strand board finished with exterior-grade resin and primed textured finish. Manufacturer claims formaldehyde-free adhesive and one-half linear expansion of hardboard. Lap siding comes in 6-in., 8-in., 9½-in., and 12-in. widths, in ⁷⁄₁₆-in. thickness, and in 12-ft., and 16-ft. lengths. Products are dimensionally stable and provide structural shear strength in exterior walls.

## 296 Masonite Corp.

1 S. Wacker Dr., Chicago, IL 60606
Phone: 800 255 0785 or 312 750 0900
Fax: 312 750 0958

**Price Index Number:** **0.5**
**Price Index Standard:** **redwood lap siding**

### HARDBOARD SIDING

Engineered hardboard residential lap and shingle siding. Products come in $7/16$-in. (East Coast only) and $1/2$-in. thicknesses and varying widths in 16-ft. lengths. Factory finished in various colors (East Coast only) with smooth and wood-grain textures. Siding is 50% denser than wood and resists splitting and checking. Installation must allow for thermal expansion of siding along its length.

## 297 Nascor, Inc.

1212 34th Ave. S.E., Calgary, AB, Canada T2G 1V7
Phone: 403 243 8919    Fax: 403 243 3417

**Price Index Number:** **0.90**
**Price Index Standard:** **residential stick framing**

### MANUFACTURED ROOFING AND SIDING

A patented framing system of wood I-joists, studs, and EPS panels on 24 in. o.c. for walls. Wall framing is $5\frac{1}{2}$ in. or $7\frac{1}{4}$ in. thick. Primary market is residential construction in Montana and Germany.

## 298 Resource Conservation Structures, Inc.

530 8th Ave. S.W., Suite 1000,
Calgary, AB, Canada T2P 3S8
Phone: 403 264 4928    Fax: 403 266 6365

**Price Index Number:** **1.05-1.1**
**Price Index Standard:** **residential stick framing**

### MANUFACTURED WALL PANELS

STRUCTURAL INSULATED PANEL is a sandwich panel of OSB and EPS in 8-ft. by 24-ft. sizes.

## 299 Shelter Enterprises

PO Box 618, Saratoga St., Cohoes, NY 12047
Phone: 800 836 0719 or 518 237 4101
Fax: 518 237 0125

**Price Index Number:** **0.90**
**Price Index Standard:** **2x6 framed wall with insulation**

### MANUFACTURED WALL PANELS

EPS core between builder's choice of stress skins, such as OSB, plywood, gypsum, and kraft paper. For exterior walls. Comes in 4-ft. by 9-ft. through 16-ft. sizes with thicknesses of $3\frac{1}{2}$ in., $5\frac{1}{2}$ in., $7\frac{1}{2}$ in., and $9\frac{1}{2}$ in. Custom panels can be made to specifications. R-value of R-23 for $5\frac{1}{2}$-in. core panel with plywood skins. Guaranteed stable R-value. Available with spline system for roof and wall residential post-and-beam construction. No CFCs used in manufacture or in final product. Typical distribution is east of the Mississippi River.

## 300 Smurfit Newsprint Corp.

427 Main St., Oregon City, OR 97045
Phone: 800 547 6633 or 503 650 4274
Fax: 503 650 4519

**Price Index Number:** **0.20**
**Price Index Standard:** **redwood clear heart
lap siding**

### HARDBOARD SIDING
CLADWOOD exterior siding is a resin-bonded particleboard substrate with a fiber overlay on both sides. Available in 4-ft. by 8-ft. and 9-ft. sheets or in shiplap 48⅜ in. by 96 in. and 48⅜ in. by 108 in., in thicknesses of ⁷⁄₁₆ in. or ½ in. Product requires a vapor barrier of one perm or less on the warm side of the wall. Comes with a 20-year warranty.

## 301 U.S. Building Panels, Inc.

10901 Lakeview Ave. S.W., Tacoma, WA 98499
Phone: 206 581 0288   Fax: 206 581 0344

**Price Index Number:** **1.0**
**Price Index Standard:** **residential stick
framing with batts**

### MANUFACTURED WALL PANELS
EPS core between OSB Sturdi-Wood. No CFCs or formaldehyde in final product. For exterior floors, roofs, and walls. Assembled with splines and screws onto residential post-and-beam structures. R-value is R-23.6 for 6-in. panel and R-29.5 for 8 in. Comes with predrilled wiring chases. Available in 4-ft. by 8-in., 10-in., and 12-in. thicknesses up to 24 ft. in length. Manufacturer assembles a demonstration house at factory location once a month.

## 302 Werzalit of America

PO Box 373, Bradford, PA 16701
Phone: 800 999 3730 or 814 362 3881
Fax: 814 362 4237

**Price Index Number:** **1.0**
**Price Index Standard:** **cedar primed
siding, grade A**

### HARDBOARD SIDING
Very durable cladding of a thermoset acrylic coating laminated to a shredded hardwood board in a compression-molded process. Prepunched, tongue-and-groove shiplap siding for new or retrofit applications. Available in 12-ft. by 4-in., 6-in., or 8-in. sizes and factory painted with custom-order colors. Painted finish comes with a 15-year warranty, material with a 20-year warranty.

## 303 Wolverine Technologies

17199 Laurel Park Dr., Suite 201,
Livonia, MI 48152-2679
Phone: 800 521 9020 or 313 953 1100
Fax: 313 953 9070

**Price Index Number:** **0.20**
**Price Index Standard:** **redwood shiplap
siding, grade A**

### VINYL SIDING
WEATHERSTONE is a vinyl siding made from pre-consumer recycled vinyl backing combined with a virgin PVC outer layer. Available in standard lap siding sizes in 7 colors. Comes with a 50-year transferable warranty.

### 304 Marlite

202 Harger St., Dover, OH 44622
Phone: 330 343 6621
Fax: 330 343 7296 or 330 343 4668 (order)

**Price Index Number:** 4.0
**Price Index Standard:** gypsum board

**FIBERBOARD**
MARLITE is a plank wall covering made from a high-percentage recycled wood-fiber substrate with a melamine topcoat. More durable than drywall, it's intended for high-traffic, commercial interiors. Available in sheets 16 in. by 96 in. by ¼ in. with a tongue-and-groove joinery system in 35 finishes, which include various colors and wood-grain patterns. Fire rating is Class C, with MARLITE FIRETEST plank Class A.

### 305 Phenix Biocomposites, Inc.

PO Box 609, Mankato, MN 56002-0609
Phone: 800 324 8187 or 507 931 9787
Fax: 507 931 5573

**Price Index Number:** 0.70
**Price Index Standard:** Corian

**FIBERBOARD**
ENVIRON is a composite building material manufactured from 40% soybean flour, 40% waste cellulose (newspaper), and 20% coloring with miscellaneous waxes and oils. Designed for use as paneling, flooring, molding, and in cabinets and furniture. Available in 10 colors in 3-ft. by 6-ft. sheets in ¼-in. increments up to 1 in. thick. Can be worked with standard woodworking tools. The factory smells more like a bakery with the aroma of roasted soybeans.

### 306 Bedard Cascades, Inc.

351 Rue Alice, Joliette, QC, Canada J6E 8P2
Phone: 514 756 4077   Fax: 514 756 4952

**Price Index Number:** 0.80
**Price Index Standard:** fiberglass roofing felt

**ROOFING FELT**
Organic roofing felt manufactured from asphalt, 20% to 30% recycled paper, and 70% recycled wood pulp.

### 307 GS Roofing Products Co., Inc.

5525 MacArthur, #900, Irving, TX 75038
Phone: 800 999 5150 or 972 580 5604
Fax: 800 999 5350

**Price Index Number:** 0.80
**Price Index Standard:** fiberglass roofing felt

**ROOFING FELT**
Organic roofing felt manufactured from asphalt, 100% recycled paper, and sawdust chips.

### 308 Innovative Formulation Corp.

1810 S. Sixth Ave., Tucson, AZ 85713-4608
Phone: 800 346 7265 or 520 628 1553
Fax: 520 628 1580

**Price Index Number:** 1.7
**Price Index Standard:** hot mop tar

**POLYESTER ROOFING**
MIRRORSEAL is combined with MIRRORFAB to form a continuous filament, single-ply, water-based, fluid-applied roofing system for flat roofs. Applied with a paint roller with no heat or smell. Can be used for roofs, decks, pools, or shower pans. Has both insulative and reflective qualities. Noncombustible and withstands extreme temperatures while maintaining flexibility. Base-coat paint is environmentally friendly, even edible, and made with salt water. White top-coat paint contains titanium oxide. Installation video available. Has a 5-year warranty with optional additional 5-year extended warranty. Free freight with minimum $1,000 order.

## 309 Bonded Fiber Products, Inc.

2748 Tanager Ave., Commerce, CA 90040
Phone: 213 726 7820   Fax: 213 726 2805

**Price Index Number:** 1.0
**Price Index Standard:** virgin polyester roofing felt

**ROOFING FELT**
QuLine roofing fabric made from 90% recycled PET plastics from 2-liter soda bottles. Only manufacturer of recycled roofing fabric in the United States.

## 310 CertainTeed Corp.

PO Box 860, Valley Forge, PA 19482
Phone: 800 274 8530 or 800 328 5874
Fax: 313 953 0855

**Price Index Number:** 0.80
**Price Index Standard:** fiberglass roofing felt

**ROOFING FELT**
Organic felt-based shingles incorporating recycled paper and recycled waste slag. Made from 20% to 25% recycled contents. Product weighs 215 to 230 lbs. per square. Comes with a 20- to 30-year warranty.

## 311 Tamko Roofing Products, Inc.

PO Box 1404, Joplin, MO 64802-1404
Phone: 800 641 4691 or 417 624 6644
Fax: 417 624 8935

**Price Index Number:** 0.80
**Price Index Standard:** fiberglass roofing felt

**ROOFING FELT**
Organic roofing felt manufactured from recycled paper, recycled wood pulp, and asphalt. Product specification number is ASTM-226 Type I or Type II.

## 312 Bass and Hays Foundry, Inc.

PO Box 531610, 238 South Bagdad Rd.,
Grand Praire, TX 75053
Phone: 800 258 2278 or 214 263 1360
Fax: 214 263 0091

**Price Index Number:** 3.0
**Price Index Standard:** sheet-metal leader

**GUTTERS, LEADERS, AND DRAINS**
Manufacturer of downspout boots, leaders, and area drains from 100% recycled cast iron.

## 313 Sun Pipe Co., Inc.

PO Box 2223, Northbrook, IL 60065
Phone: 847 272 6977   Fax: 847 272 6972

**Price Index Number:** 3.0, material/
0.5, material + labor
**Price Index Standard:** standard domed skylight

**SKYLIGHT**
SUN PIPE 13-in.- or 21-in.-dia. aluminum pipe, lined with highly reflective film (3M product), fits between rafters and joists. Brings natural light down through attic into living areas of home. Equivalent to 300 to 600 watts of light on a cloudy day, to 1,500 to 3,000 watts maximum.

## 314 AFM Enterprises, Inc.

350 W. Ash St., Suite 700, San Diego, CA 92101
Phone: 619 239 0321    Fax: 619 239 0565

**Price Index Number:** **1.3**
**Price Index Standard:** **silicon caulk**

**CAULK AND PUTTY**
SAFECOAT natural-based caulks, putties, and sealers to minimize exposure to toxic chemicals. Minimum order of 50 gallons. Products developed specifically for chemically sensitive people in consultation with environmental medicine physicians.

## 315 Auro-Sinan Co.

PO Box 857, Davis, CA 95617
Phone: 916 753 3104    Fax: 916 753 3104

**Price Index Number:** **2.0**
**Price Index Standard:** **standard ingredients**

**CAULK AND PUTTY**
AURO natural-based caulks, putties, and sealers. Hole and insulation filler made from cork, damar resin, citrus terpene, coconut oil, and natural latex. No petroleums or plastic ingredients.

## 316 Resource Conservation Tech.

2633 N. Calvert St., Baltimore, MD 21218
Phone: 410 366 1146    Fax: 410 366 1202

**Price Index Number:** **1.0-3.0**
**Price Index Standard:** **standard building gaskets**

**CAULK AND GASKETS**
Distributor of building gaskets manufactured in Sweden from cellular EPDM for superior flexibility and sealing to eliminate the use of caulks, foams, and other sealants. Variety of gasket shapes and sizes available. Distributor also claims to have better polyurethane and silicone caulks and caulk gun applicators than are commonly available.

## 317 AFM Enterprises, Inc.

350 W. Ash St., Suite 700, San Diego, CA 92101
Phone: 619 239 0321    Fax: 619 239 0565

**Price Index Number:** **1.2**
**Price Index Standard:** **chlorinated solvent**

**DUCT MASTIC**
SAFECOAT DINOFLEX joint-sealant treatment for HVAC ducts to eliminate toxic outgassing from standard sealants. Also useful as a low-toxicity roof coating to replace tar and gravel. Can be walked on and remains flexible. Products developed specifically for chemically sensitive people in consultation with environmental medicine physicians.

## 318 United McGill Corp.

PO Box 820, Columbus, OH 43216-0820
Phone: 800 624 5535 or 614 443 5520
Fax: 614 444 0234

**Price Index Number:** **0.75**
**Price Index Standard:** **chlorinated solvent**

**DUCT MASTIC**
SEAL-N-SAVE duct mastic is a low-VOC, water-based residential duct sealant applied with a caulking gun or brush for both old and new construction. UNI-MASTIC, similar to SEAL-N-SAVE, is a heavier-duty, slightly thicker mastic in gray with a higher fiber content for commercial projects.

## 319  Challenge Door Co.

902 Highway 19, Sulphur Springs, TX 75482
Phone: 800 527 1127
Fax: 800 468 3899 or 903 885 9137

**Price Index Number:  1.50**
**Price Index Standard:  solid-core wood door**

**STEEL ENTRY DOORS**
Manufacturer of 24-gauge steel residential entry doors with an expanded core of polyurethane. Comes with a 25-year warranty.

## 320  Loewen Windows

77 PTH 52 West, Steinbach, MB, Canada ROA 2AO
Phone: 800 563 9367 or 204 326 6446
Fax: 800 563 9361

**Price Index Number:  1.0, basic door**
**Price Index Standard:  solid-core wood door**

**STEEL ENTRY DOORS**
LOEWEN prehung steel door with a polyurethane core, with optional fiberglass veneer. Comes with a 25-year limited warranty.

## 321  Taylor Building Products

631 N. First St., West Branch, MI 48661
Phone: 800 248 3600 or 517 345 5110
Fax: 517 345 5116

**Price Index Number:  1.1**
**Price Index Standard:  solid-core wood door**

**STEEL ENTRY DOORS**
Manufacturer of ENERGY SAVER steel residential entry door with an expanded core of polystyrene. Pentane is used as the insulation blowing agent, which does not deplete atmospheric ozone. Available with a vinyl face and wood-grain texture.

## 322  Caridon Peachtree Doors and Windows, Inc.

4350 Peachtree Industrial Blvd.,
Norcross, GA 30071
Phone: 800 732 2499 or 770 497 2000
Fax: 800 395 3253

**Price Index Number:  0.7**
**Price Index Standard:  solid-core wood door, prehung**

**COMPOSITE DOORS**
Polyurethane core doors with composite or galvanized steel skins. Lifetime guarantee. Durable and virtually maintenance free.

## 323  Jeld-Wen

3250 Lakeport, Klamath Falls, OR 97601-0268
Phone: 800 877 9482 or 800 535 3936
Fax: 503 222 1542

**Price Index Number:  0.65**
**Price Index Standard:  solid-core wood door**

**WOOD DOORS**
ELITE EXTERIOR doors are made from molded wood fibers bonded together with heat and pressure. Composite wood-fiber skins ⅛ in. thick are molded over different frames made of medium-density fiberboard, solid-core particleboard, and laminated veneer with polystyrene core. Available in a variety of finish textures. Factory primed for painting. Doors resist warping and delaminating and come with a 5-year warranty.

## 324 Masonite Corp.

1 S. Wacker Dr., Chicago, IL 60606
Phone: 800 446 1649 (door hotline)
Fax: 601 428 1170

**Price Index Number:** **0.25, hollow core/ 0.35, solid core**
**Price Index Standard:** **solid-core fir door**

### WOOD DOORS
CRAFTMASTER molded wood fiber (hardboard) doors for interior use. Four panel designs are available with a wood-grain pattern. Paint or stain can be applied as a finish coat. The door skins are manufactured by Masonite and are then assembled by other companies. The assembled door can have finger joints and a honeycomb core.

## 325 Norco Window Co.

621 Washington St. South, Twin Falls, ID 83301
Phone: 800 526 3532   Fax: 800 526 3531

**Price Index Number:** **1.0**
**Price Index Standard:** **Andersen composite glazed doors**

### WOOD DOORS
Manufacturer of glazed wood doors with Timberstand rails and stiles.

## 326 Pease Industries

7100 Dixie Highway, Fairfield, OH 45014
Phone: 800 883 6677 or 513 870 3680
Fax: 513 870 3672

**Price Index Number:** **2.0+**
**Price Index Standard:** **solid-core wood door**

### COMPOSITE DOORS
Polyurethane core doors with fiberglass skins. HCFC-14lb is used as a blowing agent. Stainable and nearly maintenance free.

## 327 Therma-Tru Corp.

1684 Woodland Dr., Toledo, OH 43537
Phone: 800 537 8827 or 419 537 1931
Fax: 800 322 8688

**Price Index Number:** **2.0+**
**Price Index Standard:** **solid-core wood door**

### COMPOSITE DOORS
Molded fiberglass doors with polyurethane core. HCFC-14lb is used as a blowing agent. Various styles available with sidelights and textured wood-grain pattern that can be painted or stained. Comes with 25-year limited warranty.

## PLASTIC DOORS
08220

## 328 Corrim Co.

3331 County Rd. A, Oshkosh, WI 54901
Phone: 414 231 2000   Fax: 800 345 9087

**Price Index Number:** **3.0+**
**Price Index Standard:** **solid-core wood door**

### FIBERGLASS DOORS
Fiberglass-reinforced, plastic exterior doors with foamed polyurethane or balsa wood cores. Stiles and rails made from solid fiberglass-reinforced plastic. Comes with factory-applied gel coat impregnated colors, but also paintable on site. POLYFIRE is a fire-rated version of the same door, but with a mineral board core. Doors come with a 10-year warranty against corrosion and 1-year warranty for workmanship.

## 329 Premdoor Entry Systems

911 E. Jefferson St., PO Box 76,
Pittsburg, KS 66762
Phone: 800 835 0364 or 316 231 8200
Fax: 316 231 8239

**Price Index Number:  2.0+**
**Price Index Standard:  wood door**

**FIBERGLASS DOORS**
CASTLEGATE ENTRY SYSTEMS fiberglass or galvanized steel skins on exterior doors with foamed polyurethane cores. Stiles and rails with wood-grain pattern. Available in one color, but is stainable or paintable. Door panel comes with a 30-year warranty.

## 330 Visionwall Technologies, Inc.

404 123rd Ave., Suite 110,
Edmonton, AB, Canada T5V 1B4
Phone: 403 451 4000   Fax: 403 451 4745

**Price Index Number:  1.8**
**Price Index Standard:  Andersen clad
                        casement, 2' 6" x 4'**

**ALUMINUM WINDOWS**
VISIONWALL is a thermally broken aluminum casement window with two panes of Cardinal Low-E$^2$ film (PET) between two layers of clear glass. Air spaces are filled with air—no argon—and manufacturer claims it has the highest overall R-value on the American market, about R-6.3. Currently applying for NFRC rating.

## 331 Alaska Window Co.

PO Box 61252, Fairbanks, AK 99706
Phone: 800 478 5878 (AK only) or 907 479 5874
Fax: 907 479 7996

**Price Index Number:  1.35**
**Price Index Standard:  Andersen clad
                        casement, 2' 6" x 4'**

**VINYL WINDOWS**
Built to withstand the Alaskan winters, the ALASKA WINDOW has an overall R-value of just less than R-5. The glass is deeply recessed into the frame to insulate the metal edge spacer and prevent condensation. Constructed of PVC with a steel reinforcment. Casements swing inward and also tilt-in in the European fashion.

## 332 Alenco

PO Box 3309, 651 Carson St., Bryan, TX 77801
Phone: 800 528 1430 or 404 779 7770
Fax: 409 822 3259

**Price Index Number:  0.69**
**Price Index Standard:  Andersen clad
                        casement, 2' 6" x 4'**

**VINYL WINDOWS**
1500 Series vinyl windows include single hung, slider, casement, and fixed with an overall width of ¾ in. Glazing available in double pane with warm edge spaces, and Cardinal Low-E$^2$ glass. Integral vinyl color does not require painting. NFRC rated.

## 333 Alside

PO Box 2010, Akron, OH 44309
Phone: 800 257 4335 or 330 922 2202
Fax: 330 922 2354

**Price Index Number:** 0.70
**Price Index Standard:** Andersen clad
casement, 2' 6" x 4'

### VINYL WINDOWS

Vinyl window with aluminum reinforcment. Available in several styles and colors including beige, white, and wood-grain finishes. A ⅞-in. insulating glass unit is available. Integrated color never wears off.

## 334 Andersen Windows

100 4th Ave. N., Bayport, MN 55003-1096
Phone: 612 439 5150   Fax: 612 430 5214

**Price Index Number:** 1.3-1.5
**Price Index Standard:** double-glazed
aluminum-framed
casement

### WOOD WINDOWS

Manufacturer of vinyl-clad wood windows for high-end residential applications. Finger-jointed wood used in vinyl-clad frames. NFRC rated.

## 335 Caradco

201 Evans Rd., Rantoul, IL 61866
Phone: 217 893 4444
Fax: 800 225 9598 or 217 893 7595 (tech)

**Price Index Number:** 0.67
**Price Index Standard:** Andersen clad
casement, 2' 6" x 4'

### WOOD WINDOWS

Manufacturer of wood substitute and rolled aluminum-clad windows for residential applications. Wood window uses Timberstand frame. NFRC rating for wood casement using Cardinal Low-E$^2$ glass, argon filled, produces a U-value of U-0.47. Has 1-year warranty on wood, 10- to 20-year warranty on glass seal.

## 336 Champagne Industries, Inc.

12775 E. 38th Ave., Denver, CO 80329
Phone: 303 375 0570   Fax: 303 375 1212

**Price Index Number:** 0.45
**Price Index Standard:** Andersen clad
casement, 2' 6" x 4'

### VINYL WINDOWS

Vinyl windows, including single hung, slider, and fixed with an overall width of ¾ in. Glazing available in double-pane and Cardinal Low-E$^2$ glass. Integral vinyl color does not require painting. NFRC rated U-value is U-0.34. Manufacturer will offer a selection of colors and a casement window in the near future. Windows are purchased directly from the company.

## 337 Eagle Windows

375 E. 9th, Dubuque, IA 52001
Phone: 800 453 3633 or 319 556 2270
Fax: 319 556 3825

**Price Index Number:** 1.0
**Price Index Standard:** Andersen clad
casement, 2' 6" x 4'

### WOOD WINDOWS

Manufacturer of wood and aluminum-clad windows for residential applications. Wood window uses an LVL frame. NFRC rating is U-0.32 for wood and U-0.40 for aluminum, 2-ft. 6-in. by 4-ft. casement.

### 338  Gentek Building Products, Inc.

5455 E. La Palma Ave., Suite A
Anaheim, CA 92807-2922
Phone: 800 992 5226 or 714 693 0604
Fax: 714 693 0612

**Price Index Number:**     **0.70**
**Price Index Standard:**   **Andersen clad
                            casement, 2' 6" x 4'**

**VINYL WINDOWS**
ADVANTAGE COLLECTION vinyl windows and doors are
manufactured from a heavy-duty vinyl in 3 colors. Energy
conserving and low maintenance. Comes with a lifetime
limited nonprorated warranty.

---

### 339  Hurd Millworks

575 S. Whelan Ave., Medford, WI 54451
Phone: 800 433 4873 or 715 748 2011
Fax: 715 748 6043

**Price Index Number:**     **0.74**
**Price Index Standard:**   **Andersen clad
                            casement, 2' 6" x 4'**

**WOOD WINDOWS**
Manufacturer of wood and extruded aluminum clad windows
for high-end residential applications. Wood window uses
some finger-jointed frame parts with a laminated veneer for
stain-grade interior finishes. Aluminum-clad window is about
10% more expensive than the wood window. NFRC rating for
wood casement using Heat Mirror-TC88 glass produces a
U-value of U-0.26, and with Insul-8 a U-value of U-0.20.
SC-75 glazing is available for blocking excess incoming
radiation.

---

### 340  Kolbe and Kolbe Millwork Co.

1323 South 11th Ave., Wausau, WI 54401
Phone: 715 842 5666   Fax: 715 842 2863

**Price Index Number:**     **1.03**
**Price Index Standard:**   **Andersen clad
                            casement, 2' 6" x 4'**

**WOOD WINDOWS**
Manufacturer of wood windows and vinyl or aluminum-
clad windows. NFRC rating as low as U-0.30 with Cardinal
Low-E$^2$ glazing.

---

### 341  Linford Brothers Glass Co.

1245 S. 700 West, Salt Lake City, UT 84110
Phone: 800 998 3896 or 801 972 6161
Fax: 801 972 5256

**Price Index Number:**     **0.58**
**Price Index Standard:**   **Andersen clad
                            casement, 2' 6" x 4'**

**VINYL WINDOWS**
LINFORD vinyl windows are manufactured from
unplasticized vinyl chloride for stability, hardness, and
resistance to deterioration from UV light. A minimum TIO-2
ratio of 10:1 is maintained in vinyl for effective UV
resistance. NFRC rated with a maximum U-value of U-0.5.
Available in casement, single hung and fixed with a true
divided light option. Has a 20-year warranty.

---

### 342  Loewen Windows

77 PTH 52 West, Steinbach, MB, Canada R0A 2A0
Phone: 800 563 9367 or 204 326 6446
Fax: 800 563 9361

**Price Index Number:**     **1.1-1.4**
**Price Index Standard:**   **Andersen clad
                            casement, 2' 6" x 4'**

**WOOD WINDOWS**
LOEWEN HEAT-SMART extruded aluminum-clad wood
windows. Has a 10-year limited warranty.

## 343 Marvin Windows and Door Co.

PO Box 100, Warroad, MN 56763
Phone: 800 346 5044 or 218 386 1430
Fax: 218 386 1913

**Price Index Number:** 1.0
**Price Index Standard:** Andersen clad
casement, 2' 6" x 4'

### COMPOSITE WINDOW

Manufacturer of INTEGRITY line of fiberglass frame windows with interior wood cladding. Product includes casement, awning, and fixed windows with Cardinal Low-E$^2$ glass. NFRC rated with casement U-value of U-0.47 to U-0.3

## 344 Milgard

3800 136th St. N.E., Marysville, WA 98271
Phone: 800 562 8444, 206 659 0836
or 800 552 0402 (WA)
Fax: 206 922 3983 or 360 653 7229 (WA)

**Price Index Number:** 0.59
**Price Index Standard:** Andersen clad
casement, 2' 6" x 4'

### VINYL WINDOWS

Vinyl windows, sliding doors, and skylights. Both standard and custom sizes available. Some recovered material used in vinyl extrusion process. Glazing available in double pane, Cardinal Low-E$^2$ glass, or argon gas between panes. Integral vinyl color does not require painting. NFRC rating for vinyl casement with Cardinal Low-E$^2$ glazing is U-0.36.

## 345 Norco Window Co.

621 Washington St. S., Twin Falls, ID 93301
Phone: 800 526 3532   Fax: 800 526 3531

**Price Index Number:** 0.80
**Price Index Standard:** Andersen clad
casement, 2' 6" x 4'

   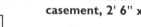

### WOOD WINDOWS

TETON Series wood windows are double glazed with Cardinal Low-E$^2$ and use finger-jointed material where not visible. SITELINE is the economy line of wood windows. Also manufacturer of aluminum extrusion-clad wood and vinyl windows for residential applications. NFRC rated.

## 346 Norco Window Co.

621 Washington St. S., Twin Falls, ID 93301
Phone: 800 526 3532   Fax: 800 526 3531

**Price Index Number:** 0.80
**Price Index Standard:** Andersen clad
casement, 2' 6" x 4'

### VINYL WINDOWS

SIERRA vinyl windows are double glazed with Cardinal Low-E$^2$. Also manufacturer of wood and aluminum extrusion-clad wood windows for residential applications. NFRC rated.

## 347 Owens Corning

One Fiberglass Tower, Toledo, OH 43659
Phone: 800 366 4430 or 614 321 7731
Fax: 419 248 1721 or 614 321 5606

**Price Index Number:** 1.21
**Price Index Standard:** Andersen clad
casement, 2' 6" x 4'

### FIBERGLASS WINDOWS

Complete line of FIBRON fiberglass frame windows filled with fiberglass insulation. Glazing available in double pane, Cardinal Low-E$^2$ glass or argon gas between panes. Frames will not rot, warp, shrink, check, or swell and can be painted. Expanding westward with distribution currently limited to areas east of the Mississippi. NFRC rated U-value for standard double glazed casement is U-0.42. Lifetime warranty on unit with 10 years on glass. Many accessories available with window.

## 348 Pella Corporation

102 Main St., Pella, IA 50219
Phone: 515 628 1000 or 515 628 6505
Fax: 515 628 6070 or 515 628 6472

**Price Index Number:**   **0.80**
**Price Index Standard:**   **Andersen clad**
                        **casement, 2' 6" x 4'**

**WOOD WINDOWS**
Manufacturer of wood windows for residential applications. SMART SASH is a single- or double-glazed window with a removable glazing panel on the inside and an optional adjustable blind located between the inside and outside glazing. Cardinal Low-$E^2$ glazing also available. NFRC-rated U-value with INSULSHEILD is U-0.23

## 349 Weather Shield Manufacturing, Inc.

PO Box 309, Medford, WI 54451
Phone: 800 222 2995   Fax: 800 222 2146

**Price Index Number:**   **0.83**
**Price Index Standard:**   **Andersen clad**
                        **casement, 2' 6" x 4'**

**VINYL WINDOWS**
PROSHIELD vinyl windows and WEATHER SHIELD vinyl-clad wood windows are manufactured from unplasticized vinyl chloride for stability, hardness, and resistance to deterioration from UV light. Lifetime warranty.

## WOOD WINDOWS      08610

## 350 Wenco Windows

PO Box 1248, Mt. Vernon, OH 43050-8248
Phone: 800 458 9128 or 614 397 1144
Fax: 800 892 8519

**Price Index Number:**   **0.60**
**Price Index Standard:**   **Andersen clad**
                        **casement, 2' 6" x 4'**

**COMPOSITE WINDOW**
ELIMINATOR-PF is a wood window with a wood fiber and resin composite sill using Werzalit with insulating glass by UltraGlass. Clad or factory finished exterior in white, beige, or umber. Wenco has manufacturing plants throughout the United States, reducing transportation costs.

## HARDWARE      08700

## 351 Resource Conservation Tech.

2633 N. Calvert St., Baltimore, MD 21218
Phone: 401 366 1146   Fax: 410 366 1202

**Price Index Number:**   **1.0-3.0**
**Price Index Standard:**   **standard building**
                        **components**

**ACCESSORIES**
Silicone weather-sealant systems, adjustable shim screws, glazing gaskets, unique door-locking system.

## 352 Van Duerr Industries

426 Broadway, Suite 207, Chico, CA 95928
Phone: 800 497 2003 or 916 893 1596
Fax: 916 893 1560

**Price Index Number:**   **0.25-0.35, material +**
                        **labor**
**Price Index Standard:**   **concrete or**
                        **aluminum ramp**

**RAMP**
EZ EDGE THRESHOLD RAMP is manufactured from 100% recycled tire rubber to provide disabled access over thresholds up to 1½ in. high, maintaining a slope of 1:12.

## 353 Southwall Technologies

1029 Corporation Way, Palo Alto, CA 94303
Phone: 800 365 8794 or 415 962 9111
Fax: 415 967 8713

**Price Index Number:**     1.15, window/
                            1.20, door
**Price Index Standard:**   double-glazed
                            windows and doors

### INSULATING GLASS

HEAT MIRROR is a polymer film covered with a molecular layer of metal. This coating is designed to selectively allow visible light to pass through while reflecting the near UV and infrared radiation. Window manufacturers assemble the HEAT MIRROR glass unit by suspending and stretching the HEAT MIRROR film in an air space between two sheets of glass using an oven heating technique. Available in different glass types and colors, and in 6 thermal performance levels. As an aesthetic consideration, the greater the thermal performance, the more pronounced is the reflective metal coating in a well-lighted space at night, resulting in a mirrored surface. Other Southwall Technologies products include HEATSEAL THERMAL BREAK SPACER and XIR SOLAR CONTROL WINDOW FILM, a clear retrofit window film product designed to block the near UV and infrared without blocking visible light.

## 354 Auro-Sinan Co.

PO Box 857, Davis, CA 95617
Phone: 916 753 3104    Fax: 916 753 3104

**Price Index Number:**     2.0
**Price Index Standard:**   standard joint
                            compound

### PLASTER/JOINT COMPOUND

AURO joint cement and texture compounds (product #311) formulated with inert fillers and natural binders. Not intended for sheetrock joints, but as a covering surface. No petroleums or plastic ingredients. Manufacturer claims this product to be nontoxic.

## 355 GP Gypsum Corp.

133 Peachtree St. N.E., Atlanta, GA 30303
Phone: 800 947 4497 (sales) or
800 225 6119 (tech)    Fax: 404 588 5157 (tech)

**Price Index Number:**     1.0
**Price Index Standard:**   gypsum board

### GYPSUM BOARD

GYPROC board contains up to 10% recycled scrap wallboard and up to 100% by-product gypsum. Facing paper is recycled newspaper and corrugated cardboard. Available in thicknesses of ½ in. and ⅝ in. and widths of 48 in. and 54 in.

## 356 Louisiana-Pacific

111 S.W. Fifth Ave., Portland, OR 97204
Phone: 800 999 9105, 800 547 6331
or 503 221 0800    Fax: 503 796 0204

**Price Index Number:**     2.0-2.5
**Price Index Standard:**   gypsum board

### GYPSUM BOARD

FIBERBOND and FIBERBOND VHI are fiber-reinforced gypsum wallboards manufactured from recycled cellulose fiber from telephone books and newspapers, gypsum, and perlite (an expanded mineral aggregate). Designed to outperform conventional gypsum wallboard. Can be nailed, screwed, and stapled, providing uniform strength and superior holding power. Available in both ½-in. and ⅝-in. thicknesses. Class X fire rated for 45 minutes. Recycled content is 30%. FIBERBOND SHEATHING is the same as FIBERBOND with a silicate coating for exterior use as a sheathing and substrate for synthetic stucco. Shipping cost due to location of manufacturing plant in Nova Scotia.

### 357 Murco Wall Products, Inc.

300 N.E. 21st St., Fort Worth, TX 76106
Phone: 800 446 7124 or 817 626 1987
Fax: 817 626 0821

**Price Index Number:** **1.0**
**Price Index Standard:** **standard joint compound**

## GYPSUM BOARD ACCESSORIES 09270

### 358 The Millennium Group

121 S. Monroe, Waterloo, WI 53594
Phone: 800 280 2304, 800 504 0043
or 815 385 3145 Fax: 414 478 9630

**Price Index Number:** **0.33, 5 clips**
**Price Index Standard:** **wood stud**

**GYPSUM BOARD CLIP**
THE NAILER is a drywall clip manufactured from recycled plastic (HDPE). This product reduces the use of studs in a framed wall while improving thermal efficiency. Manufacturer claims labor and material savings on a 2,700-sq.-ft. house to be $200 to $300 with the elimination of 30 to 35 studs and 180 linear feet of backing. This product also provides a solution to truss uplift.

## TILE 09300

### 359 AFM Enterprises, Inc.

350 W. Ash St., Suite 700, San Diego, CA 92101
Phone: 619 239 0321 Fax: 619 239 0565

**Price Index Number:** **1.3**
**Price Index Standard:** **chlorinated solvent mastic**

**ADHESIVES**
SAFECOAT 3-IN-1 water-based, natural ceramic, resilient, and wood laminate floor adhesive. SAFECOAT ALMIGHTY water-based, natural carpet pad adhesive. Products meet strictest VOC environmental regulations. Products developed specifically for chemically sensitive people in consultation with environmental medicine physicians. Manufacturer claims this product to be nontoxic.

### 360 Bedrock Industries

1401 W. Garfield St., Seattle, WA 98119
Phone: 206 781 7025 Fax: 206 283 0497

**Price Index Number:** **2.6, stone fragment tile**
**Price Index Standard:** **American Olean economy tile**

**RUBBLE STONE FLOORING TILE**
100% recycled stone tile fragments 1 in. to 6 in. in length and ¾ in. thick, for use in mosaic stone flooring. Stone tile fragments are tumbled to create a weathered look with rounded corners. Manufacturer claims this product to be nontoxic.

### 361 Bedrock Industries

1401 W. Garfield St., Seattle, WA 98119
Phone: 206 781 7025 Fax: 206 283 0497

**Price Index Number:** **7.2, GLAZESTONE**
**Price Index Standard:** **American Olean economy tile**

**CERAMIC TILE**
GLAZESTONE tiles are manufactured from 100% recycled glass in 30 colors plus clear in 4-in. by 4-in. sizes. Manufacturer claims this tile is very durable and will not scratch. Also available is a 9-in. by 4-in. leaf border tile. Manufacturer claims this product to be nontoxic.

## 362 Crossville Ceramics

PO Box 1168, Crossville, TN 38557
Phone: 615 484 2110   Fax: 615 484 8418

**Price Index Number:** 1.0
**Price Index Standard:** **American Olean economy tile**

### CERAMIC TILE

ECO-TILE is a tile manufactured from recycled waste tile material generated within the manufacturing plant. Available in 8 in. by 8 in. with 6-in. by 8-in. cove base and 1-in. by 6-in. inside and outside corners. The 2 surface finishes available are unpolished with sheen and crosstread with sheen. Color is gray commingled, and each batch varies depending upon availability of waste tile. Requires a 6- to 8-week order lead time because it is not a stocked item. Manufacturer claims this product to be nontoxic.

## 363 Dal-Tile Corp.

PO Box 17130, Dallas, TX 75217
Phone: 800 933 8453 or 214 398 1411
Fax: 972 788 1907

**Price Index Number:** 1.0-1.7
**Price Index Standard:** **American Olean economy tile**

### CERAMIC TILE

DURAFLOR glazed paver tile is a glass-bonded ceramic tile manufactured from reprocessed scrap tile fragments. Available in about 60 standard colors for residential floors. Sizes are 8 in. by 8 in. by $5/16$ in. and 12 in. by 12 in. by $5/16$ in. Manufacturer claims this product to be nontoxic.

## 364 Metropolitan Ceramics

PO Box 9240, Canton, OH 44711-9240
Phone: 330 484 4887   Fax: 330 484 4880

**Price Index Number:** 1.3
**Price Index Standard:** **American Olean economy tile**

### CERAMIC TILE

Quarry tile manufactured from recycled in-house waste materials. Tile #814 IRONROCK SPECIAL and #81X IRONROCK SPECIAL X, with slip-resistant textured surface from a granular iron additive. Appropriate for inside or outside use. Comes in nominal size of 8 in. by 8 in. by ½ in. This product is not part of the company's standard line of tiles and is not listed in its catalog because it is a nonstandard color. Manufacturer claims this product to be nontoxic.

## 365 Summitville Tiles, Inc.

PO Box 73, State Rt. 644, Summitville, OH 43962
Phone: 330 223 1511   Fax: 330 223 1414

**Price Index Number:** 1.15-1.7
**Price Index Standard:** **American Olean economy tile**

### CERAMIC TILE

Glazed porcelain pavers from a by-product of feldspar mining. Manufacturing system reuses 100% of solid waste. Manufacturer claims this product to be nontoxic.

## 366 Terra-Green Technologies

1650 Progress Dr., Richmond, IN 47374
Phone: 317 935 4760   Fax: 317 935 3971

**Price Index Number:** 2.0-4.2
**Price Index Standard:** **American Olean economy tile**

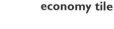

### CERAMIC TILE

TRAFFIC TILES are glass-bonded ceramic tiles made with over 70% recycled auto and aerospace industry glass. High density limits moisture absorption and provides temperature-resistant surface. Commercial grade. Available in various colors and sizes for floors, countertops, and walls. TIERRA CLASSIC is a residential designer tile with a rustic hand-molded look and manufactured from a glass-bonded ceramic with at least 55% recycled glass content. Available in 14 colors and a 4-in. by 4-in., 4-in. by 8-in., 8-in. by 8-in., 12-in. by-12 in., a 10-in. octagon, and 8-in. clipped corner tile. Intended for interior applications. Not recommended for high-traffic applications. Manufacturer claims this product to be nontoxic.

## 367 Syndesis Studio

2908 Colorado Ave., Santa Monica, CA 90404
Phone: 310 829 9932   Fax: 310 829 5641

**Price Index Number:**   **1.0**
**Price Index Standard:**   **granite**

**PRECAST TERRAZZO**
SYNDECRETE is a lightweight, precast concrete surfacing material used in tabletops, countertops, sinks, tub surrounds, tile flooring, site works, and furniture. For countertop applications, sink cutouts are made on-site. Aggregates can be added to vary the texture and appearance. Made to custom order and can utilize recycled materials including stone, wood, steel shavings, and plastic chips for unique patterns and designs. At 75 lbs./cu. ft., the weight is one-half that of standard concrete, and SYNDECRETE can be worked with standard woodworking tools. Fly ash, volcanic ash, and polypropylene fibers in the mix make it more resistant to cracking and chipping than standard concrete. Appropriate for interior or exterior use.

## 368 American Sprayed Fibers, Inc.

1550 E. 91st Dr., Merrillville, IN 46410
Phone: 800 824 2997   Fax: 219 736 6126

**Price Index Number:**   **0.33, sound transmission coefficient/ 0.50, noise reduction coefficient**
**Price Index Standard:**   **fiberglass batt**

**ACOUSTICAL INSULATION**
SOUND-PRUF, made from rock wool and cellulose, is spray-applied and uses a liquid adhesive.

## 369 Armstrong World Industries

PO Box 3001, Lancaster, PA 17604
Phone: 800 448 1405 or 717 397 0611
Fax: 717 396 2126

**Price Index Number:**   **1.0**
**Price Index Standard:**   **standard tile**

**ACOUSTICAL CEILINGS**
Ceiling tile manufactured from recycled cellulose, mineral wool, perlite, and clay. Intended for both commercial and residential projects, the tiles are available in 3 sizes: 12 in. by 12 in., 24 in. by 24 in., and 24 in. by 48 in. Although these tiles are not fire rated, they are approved for use in UL fire-rated assemblies.

## 370 Chicago Metallic

4849 S. Austin Ave., Chicago, IL 60683
Phone: 800 930 9990 or 708 563 4600
Fax: 800 222 3744

**Price Index Number:**   **1.2**
**Price Index Standard:**   **standard mineral fiber panel**

**ACOUSTICAL CEILINGS**
EUROSTONE is an acoustical ceiling panel manufactured from perlite fired and bonded together with an inorganic binder. Manufacturer claims product to be nontoxic and recyclable.

## 371 Fibrex, Inc.

PO Box 1148, Aurora, IL 60507-1148
Phone: 800 342 7391 or 708 896 4800
Fax: 708 896 3200

**Price Index Number:** 0.70
**Price Index Standard:** EPS per R-value

### ACOUSTICAL INSULATION
FBX-OEM panel insulation from mineral slag. R-value is R-4/in. Available in a variety of densities, 4.0 to 12.0 lbs./cu. ft.

## 372 Homasote Co.

PO Box 7240, West Trenton, NJ 08628-0240
Phone: 800 257 9491 or 609 883 3300
Fax: 609 530 1584

**Price Index Number:** 0.13
**Price Index Standard:** fabric over acoustical mineral board

### ACOUSTICAL INSULATION
SOUND-A-SOTE compressed cellulose fiber composition board for use beneath gypsum, made from 100% recycled newsprint. Available in 4-ft. by 8-ft. by ½-in. sheets. Treated to protect against termites, rot, and fungi. No urea formaldehyde or asbestos.

## 373 Solomit Strawboard Pty. Ltd.

26 Glomar Ct., Dandenong 3175, Victoria, Australia
Phone: 03 793 3088    Fax: 03 701 3080

**Price Index Number:** 5.0
**Price Index Standard:** standard mineral fiber panel

### ACOUSTICAL CEILINGS
Attractive, natural straw ceiling panels with decent sound absorption and thermal qualities. Comes in 1-in. and 2-in. thicknesses, 4-ft. widths, and up to 8-ft. lengths. Creates an attractive textured ceiling and meets Australian building code requirements for installation in public places. Variety of installation methods.

### WOOD FLOORING 09550

## 374 Albany Woodworks

PO Box 729, Albany, LA 70711-0729
Phone: 504 567 1155    Fax: 504 567 5150

**Price Index Number:** 1.8, antique heart pine
**Price Index Standard:** red oak flooring

### RECYCLED FLOORING
Long-leaf heart pine flooring recycled from 18th-century floors in buildings throughout the southeastern United States. Available in ¾ in. by 3 in., 4 in., 5 in. and 5¼ in., or 1½ in. by 11 in. or wider in 3 grades. Flooring is tongue-and-groove, center-matched random lengths.

## 375 Conklin's Authentic Antique Barnwood

RR 1, Box 70, Susquehanna, PA 18847
Phone: 717 465 3832    Fax: 717 465 3832

**Price Index Number:** 1.4 white pine, hem fir/2.2 heart pine
**Price Index Standard:** red oak flooring

### RECLAIMED FLOORING
Pine, hemlock, chestnut, oak, and heart pine flooring from remilled antique beams from 19th-century barns. Flooring available in random lengths and widths.

## 376 Eco Design/Natural Choice

1365 Rufina Circle, Santa Fe, NM 87505
Phone: 800 621 2591 or 505 438 3448
Fax: 505 438 0199

**Price Index Number:** **1.1-1.7**
**Price Index Standard:** **oil-based**
**polyurethanes**

### WOOD-FLOOR FINISHES
Importer and distributor of LIVOS finishes and waxes for wood floors. Products have a citrus base with an orange smell when applied. Other common ingredients include beeswax and linseed oil. Product of Germany.

## 377 EcoTimber

1020 Heinz Ave., Berkeley, CA 94710
Phone: 510 549 3000   Fax: 510 549 3001

**Price Index Number:** **1.0**
**Price Index Standard:** **non-sustainably**
**harvested sources**

### WOOD FLOORING
Wood flooring manufactured from a sustainably harvested source, including laminated Mexican cherry and walnut and Wisconsin maple. Endorsed by the Rainforest Action Network.

## 378 Goodwin Heart Pine Co.

106 S.W. 109th Pl., Micanopy, FL 32667
Phone: 800 336 3118 or 352 466 0339
Fax: 352 466 0608

**Price Index Number:** **2.0**
**Price Index Standard:** **red oak flooring**

### RECLAIMED FLOORING
Tongue-and-groove flooring, dimensional lumber, and beams milled from select vertical and curly grades of virgin heart pine logs or old-growth cypress reclaimed from riverbeds.

## 379 The Joinery Co.

PO Box 518, Tarboro, NC 27886
Phone: 800 726 7463 or 919 823 3306
Fax: 919 823 0818

**Price Index Number:** **1.2-2.9, pine/**
**0.80-2.3, Douglas fir**
**Price Index Standard:** **red oak flooring**

### RECYCLED FLOORING
Flooring remilled from timbers recycled from Early American buildings. Wood is long-leaf heart pine in select prime standard sizes of 4/4 thicknesses and in widths from 3 in. to 7 in. Custom-specified widths milled. Other custom products include fine furniture.

## 380 Kentucky Wood Floors

PO Box 33276, Louisville, KY 40232
Phone: 800 235 5235 or 502 451 6024
Fax: 502 451 6027

**Price Index Number:** **1.7**
**Price Index Standard:** **red oak flooring**

### PARQUET FLOORING
Manufacturer of custom planks and parquet hardwood flooring from any wood species and pattern specified. Available in specified sizes from 9 in. by 9 in. to 39 in. by 39 in. Parquet flooring takes advantage of small pieces of wood that might otherwise be considered scrap. Manufacturer claims this product to be nontoxic.

## 381 Rodman Industries

PO Box 76, Marinette, WI 54143
Phone: 715 735 9500   Fax: 715 735 6148

**Price Index Number:** **1.3**
**Price Index Standard:** **particleboard**

**PARTICLEBOARD**
RESINCORE I-45, -55, and -62 made from sawdust, phenolic resin, and wax is a water-resistant particleboard for nonstructural applications. RESINCORE I-62 is a 62-lb., high-density panel for industrial floor applications to replace wood strip flooring that is twice as hard as oak. Contains no formaldehyde.

## 382 Smith and Fong Co.

2121 Bryant St., #203, San Francisco, CA 94110
Phone: 415 285 4889   Fax: 415 285 8230

**Price Index Number:** **2.0**
**Price Index Standard:** **red oak flooring**

**BAMBOO FLOORING**
PLYBOO is a 100% bamboo flooring manufactured as a 3-ply laminate, tongue-and-groove strip, 3⅜ in. by 72 in. by ⅝ in. Available in natural white or amber. Bamboo has twice the stability of red oak and 90% of the hardness. Installation and maintenance is similar to a hardwood floor. As a grass, bamboo grows quickly and in abundance.

## 383 Stein and Collett, Inc.

PO Box 4065, McCall, ID 83638
Phone: 208 634 5374   Fax: 208 634 8228

**Price Index Number:** **2.0**
**Price Index Standard:** **red oak flooring**

**RECYCLED FLOORING**
Recycled heart pine wood flooring salvaged from building demolition.

## 384 Sylan and Dean Brandt

651 E. Main St., Lititz, PA 17543
Phone: 717 626 4520   Fax: 717 626 5867

**Price Index Number:** **2.7**
**Price Index Standard:** **red oak flooring**

**RECYCLED FLOORING**
Tongue-and-groove wood flooring recovered from old factories. Available in yellow pine and oak, up to 13 in. wide and 1 in. thick. Custom milled.

## 385 UGL

PO Box 70, Scranton, PA 18501
Phone: 800 845 5227 or 717 344 1202
Fax: 717 969 7634

**Price Index Number:** **1.4**
**Price Index Standard:** **oil-based polyurethanes**

**WOOD-FLOOR FINISHES**
AQUA ZAR is a water-based polyurethane containing mostly aliphatic hydrocarbons instead of the mostly carcinogenic aromatic hydrocarbons found in solvent-based finishes. If a polyurethane is to be used, this product has been found to be one of the best for chemically sensitive people, as reported by *Environmental Building News.* Manufacturer claims product to be nontoxic.

### 386 What It's Worth

PO Box 162135, Austin, TX 78716
Phone: 512 328 8837   Fax: 512 328 8837

**Price Index Number:** **1.8**
**Price Index Standard:** **red oak flooring**

**RECYCLED FLOORING**
Offset tongue-and-groove flooring and "Rustic Grade" ceiling planks remilled from timbers recycled from buildings. Wood is long-leaf heart pine in standard sizes of ¾ in. thickness, and widths of 2¼ in., 3¼ in., 4¼ in., or 5¼ in. Also available is Louisiana tidewater cypress reclaimed from 18th- and 19th-century sawn logs lodged in riverbeds.

### 387 The Wood Cellar

119 W. Perry St., Savannah, GA 31401
Phone: 800 795 9114 or 912 323 5819
Fax: 912 323 6063

**Price Index Number:** **1.2, antique pine/**
**1.6, antique oak**
**Price Index Standard:** **red oak flooring**

**RECYCLED FLOORING**
Flooring remilled from timbers recycled from late 19th-century warehouses and factories. Wood is long-leaf heart pine in standard sizes of ¾ in. thickness, and widths of 3 in. to 5 in. in random length. Also available in a long strip panel ⅝ in. by 7⅜ in. by 7 ft. 10½ in., in 2 or 3 strip styles, and parquet flooring in 12-in. squares. Custom-specified widths milled.

### 388 Wood Floors, Inc.

PO Box 1522, Orangeburg, SC 29116-1522
Phone: 803 534 8478   Fax: 803 533 0051

**Price Index Number:** **1.9, antique pine/**
**1.4, antique oak**
**Price Index Standard:** **new heart pine /**
**red oak flooring**

**RECYCLED FLOORING**
Flooring remilled from timbers recycled from Early American buildings. Wood is long-leaf heart pine in thicknesses of ¾ in. or ⅝ in. and widths from 3 in. to 8 in., 10 in., and 12 in.

### 389 Alpha Granite and Marble

2303 Merced St., San Leandro, CA 94577
Phone: 510 357 6140
Fax: 510 357 6139 or 510 357 6365

**Price Index Number:** **2.0**
**Price Index Standard:** **virgin rubber floor**

**TILE AND SHEET FLOORING**
EL DORADO VELVET TILE manufactured from recycled tire rubber and nylon cord vulcanized to rubber with a fabric reinforced backing. Tile is 12 in. by 12 in. by ⅜ in. with a chenille finish and typically installed in a basketweave pattern. Available in black/gray color. Comes with a guarantee against any surface wear for the first 5 years of installation.

### 390 Amtico International

200 Lexington Ave. #809, New York, NY 10016
Phone: 800 480 3960 or 212 545 1127
Fax: 212 545 8382

**Price Index Number:** **each project is**
**individually bid**

**VINYL FLOORING**
Custom designed and manufactured vinyl flooring with 100% recycled PVC backing. Solvent-free latex adhesive. Comes with a commercial application guarantee for 10 years.

## 391 Carlisle Tire and Rubber Co.

PO Box 99, Carlisle, PA 17013
Phone: 800 851 4746 or 717 249 1000
Fax: 717 249 0015

**Price Index Number:** **2.0**
**Price Index Standard:** **virgin rubber floor tiles, 3/16 in. thick**

### TILE FLOORING
SOFTPAVE tiles are molded from 70% to 80% recycled tires with a rubber binder. Applicable to patios, decks, walkways, and athletic-facility floors. Produced in a 24-in. by 24-in., 1-in.-thick square cut or interlocking tile. Interlocking tile comes with a nylon batten. Can be installed loose or directly adhered to a substrate of concrete, asphalt, wood, gravel, or crushed stone. Available in red, black, gray, green flecked on black, or red flecked on black.

## 392 Dinoflex Manufacturing Ltd.

PO Box 3309, Salmon Arm, BC, Canada V1E 4S1
Phone: 250 832 7780   Fax: 250 832 7788

**Price Index Number:** **0.90-1.1**
**Price Index Standard:** **virgin rubber flooring**

### MAT FLOORING
SPORT MAT are rubber floor mats manufactured from 60% to 93% recycled rubber and available in various colors. Square cut (38 in. by 38 in.) or interlocked (37½ in. by 37½ in.) tile in 5/16-in., 3/8-in., or ½-in. thicknesses. Manufactured from granulated tire rubber bound with polyurethane to create a perforated tile for drainage that is self-cleaning with a nonskid surface. Speckled color pattern added with EPDM granules. CUSHION WALK basketweave pattern arranged into a 24-in. by 24-in. by ¾-in. tile for exterior nonskid walkways. Available in brick red, green, black, and gray. DINOFLEX PLAYTILES are 3 in. thick for playground surfacing.

## 393 DLW Gerbert Ltd.

715 Fountain Ave., Lancaster, PA 17604
Phone: 800 828 9461   Fax: 717 394 1937

**Price Index Number:** **1.0, linoleum**
**Price Index Standard:** **resilient vinyl flooring**

### TILE AND SHEET FLOORING
Natural linoleum manufactured from linseed oil, pine tree resins, wood flour, and cork. Manufacturer claims product to be nontoxic.

## 394 Dodge-Regupol

PO Box 989, Lancaster, PA 17608-0989
Phone: 800 322 1923 or 717 295 3400
Fax: 717 295 3414

**Price Index Number:** **1.0, EVERLAST/ 0.70, EVEROLL**
**Price Index Standard:** **resilient vinyl flooring**

### TILE AND SHEET FLOORING
EVERLAST rubber flooring tiles from 100% recycled tires in 18-in. or 36-in. square tiles 5/32 in., ¼ in., or 3/8 in. thick, or EVEROLL rolls up to 36 in. wide by 36 ft. long. Applicable to athletic-facility floors and other such high-use areas. Available in a variety of colored patterns on a black background. SOFTSTONE is a cleaned version of the product with a gray color. Approved for use with under-floor heating systems. No evidence of toxins such as PCB, mercury or formaldehyde. Absorbs sound well.

## 395 Dodge-Regupol

PO Box 989, Lancaster, PA 17608-0989
Phone: 800 322 1923 or 717 295 3400
Fax: 717 295 3414

**Price Index Number:** **0.86**
**Price Index Standard:** **resilient vinyl flooring**

### CORK FLOORING TILES
Cork flooring tiles for commercial and residential use. Cork tiles are a soundproof underlayment for ceramic and wood floors. Comes in 12 in. by 12 in. by 3/16 in. or 5/16 in. in unfinished or prefinished polyurethane or wax coatings. The unfinished surface comes in one light color. The prefinished surfaces come in light, medium, and dark colors. Cork is a renewable resource harvested on a 9-year cycle. Cork for this product is 100% recycled from bottle-cork industry waste.

## 396 Durable Mat Co.

75 N. Pleasant St., Norwalk, OH 44857
Phone: 800 537 1603   Fax: 800 537 6287

**Price Index Number:** 1.0+
**Price Index Standard: rubber flooring**

### TILE AND SHEET FLOORING

Resilient flooring made from recycled truck tires. Suitable for interior or exterior use. DURA-TILE II flooring is suitable for commercial applications, manufactured as 12-in. by 12-in. by ⅜-in. tiles with a chenille-like textured surface. Available in assorted shades of brownish gray.

## 397 Eco Design/Natural Choice

1365 Rufina Circle, Santa Fe, NM 87505
Phone: 505 438 3448   Fax: 505 438 0199

**Price Index Number:** 0.93
**Price Index Standard: resilient vinyl flooring**

### CORK FLOORING TILES

Cork flooring tiles for commercial and residential use. Cork tiles are a soundproof underlayment for ceramic and wood floors. Comes in 11¾-in. by 11¾-in. unfinished tiles. Unlike synthetic materials, cork does not release toxic fumes. A good insulator over concrete slabs, cork is also durable and anti-static with an attractive amber color. Cork is a renewable resource harvested from the bark of cork oaks on a 9-year cycle without destroying the trees. Manufacturer claims product to be nontoxic.

## 398 Elsro, Inc.

114 37th St., Evans, CO 80620
Phone: 800 661 6971   Fax: 970 330 5123

**Price Index Number:** 0.4-0.7
**Price Index Standard: acrylic flooring, ½ in.**

### ASPHALT PLANK TILE

Commercial floor tiles for industrial use manufactured from paper, polypropylene, asphalt, and mineral fillers. Recycled newspaper and asphalt make up 20% of product. Available in black, 12-in. by 24-in. by ½-in. tiles. Treated with an antistatic finish.

## 399 Flexco Co.

PO Box 553, Tuscumbia, AL 35674
Phone: 800 633 3151 or 205 383 7474
Fax: 800 346 9075

**Price Index Number:** 2.0
**Price Index Standard: rubber floor**

### TILE AND SHEET FLOORING

FLEX-TUFT 12-in. by 12-in. or 6-in. by 12-in. tiles or 12-in. by 30-ft. rolls. Nonskid rubber flooring made from recycled commercial tires. Tiles are made from nylon cord vulcanized to rubber for a chenille surface with a fabric reinforced backing. Comes with a limited 5-year warranty. For interior or exterior commercial traffic; resistant to organic deterioration or weathering.

## 400 Forbo North America

PO Box 667, Hazelton, PA 18201
Phone: 800 233 0475 or 717 459 0771
Fax: 717 450 0258

**Price Index Number:** 1.0
**Price Index Standard: vinyl sheet or tiles**
                            **+ labor**

### SHEET FLOORING

FORBO MARMOLEUM is a natural linoleum made from linseed oil, pine resins, wood flour, and cork. These ingredients are mixed with clay, chalk fillers, and natural pigments, then bonded to a jute backing, and cured. Available in 79-in. by 105-in. rolls or 12-in. by 12-in. and 24-in. by 24-in. by ¹⁄₁₀-in.-thick tiles, lightly marbleized patterns in 36 colors. Natural linoleum floors have lasted 40 years. Manufacturer claims product to be nontoxic.

## 401 Indusol, Inc.

11 Depot St., Sutton, MA 01590
Phone: 508 865 9516   Fax: 508 865 9518

**Price Index Number:**   **0.8**
**Price Index Standard:**   **virgin rubber flooring**

### TILE FLOORING

POLY-FLOOR is a PVC tile floor assembled from 12-in.-square interlocking floor mats manufactured from recycled plastic.

## 402 KORQ, Inc.

400 E. 56th St., New York, NY 10022
Phone: 212 758 2593   Fax: 212 758 0025

**Price Index Number:**   **0.30, tile/**
**1.2, READY KORQ**
**Price Index Standard:**   **resilient vinyl flooring**

### TILE AND SHEET FLOORING

Cork products include wall and floor tiles and underlayment for ceramic tiles. Tiles come ¼-in. thick in sizes of 12 in. by 12 in., 16 in. by 16 in. and 30cm by 90cm. Available in 8 decorative designs and 2 densities. READY KORQ is a tongue-and-groove cork tile consisting of 3 plies, 30cm by 90cm by ½ in. thick, requiring very little use of adhesives. Middle ply consists of a recycled composite wood fiber. Easy installation. Can be ordered with a natural finish or ecological varnish. Also available is cork wallpaper, fabrics, and shower tiles. Wallpaper and fabric has suede texture. Naturally anti-bacterial. Cork is a renewable resource harvested on a 9-year cycle. Company is the export office for Sardinian cork producers. Manufacturer claims product to be nontoxic.

## 403 Lancaster Colony Commercial Products

PO Box 630, Columbus, OH 43216
Phone: 800 292 7260   Fax: 614 263 2857

**Price Index Number:**   **1.0**
**Price Index Standard:**   **virgin rubber flooring**

### TILE AND SHEET FLOORING

Rubber mats, runners, and interlocking heavy-duty tiles made from recycled tires. Runners are 24 in., 36 in., or 48 in. by 22 ft. by ¼ in. to ½ in. Typical applications include athletic-facility floors and walking surfaces adjacent to ice-skating rinks.

## 404 Mat Factory

760 W. 16th St., Suite E, Costa Mesa, CA 92627
Phone: 800 628 7626 or 714 645 3122
Fax: 714 645 0966

**Price Index Number:**   **2.6-3.2**
**Price Index Standard:**   **interlocking landscape**
**pavers, concrete**

### TILE FLOORING

SAFETY DECK II ground tiles manufactured from 100% recycled PVC and tire rubber featuring an interlocking "fishhook" design, requiring a special tool for disassembly. Tiles are 20 in. by 20 in. by 1 in., hollow underneath with 1-in.-square holes for grass to grow through. Applicable to both residential and commercial uses including playgrounds and handicapped access areas.

## 405 Mats, Inc.

PO Box 916, Braintree, MA 02185-0002
Phone: 800 628 7462 or 617 848 6313
Fax: 617 843 9750

**Price Index Number:**   **0.80**
**Price Index Standard:**   **virgin rubber flooring**
**+ labor**

### TILE AND SHEET FLOORING

Recycled rubber and PVC flooring products for industrial or exterior uses. ECO-TILE is an interlocking vinyl floor tile manufactured in Belgium from 100% recycled plastic. Tongue-and-groove interlocking tile with raised-disk surface, 19⅝ in. by 19⅝ in., provides monolithic, stable, loose-laid floor that can be installed with unskilled labor. No adhesives required.

## 406 Natural Cork Ltd.

1750 Peachtree St., Atlanta, GA 30309
Phone: 800 533 2675 or 800 404 2675
Fax: 404 872 4168

**Price Index Number:**  1.2
**Price Index Standard:  resilient vinyl flooring**

### TILE AND SHEET FLOORING

Cork products include wall and floor tiles and underlayment for ceramic tiles. Tiles come 12 in. by 12 in. and 12 in. by 24 in., each 5/32 in. thick. Available in 4 decorative patterns and a range of densities. PLANK PLUS is a tongue-and-groove cork tile consisting of 3 plies, 12 in. by 36 in. by 1/2 in. thick, requiring very little use of adhesives. Middle ply consists of a recycled composite wood fiber. Easy installation. Also available as a wall tile. Naturally antibacterial. Cork is a renewable resource harvested on a 9-year cycle. Company imports cork from Spain.

## 407 No Fault Industries, Inc.

11325 Pennywood Ave., Baton Rouge, LA 70809
Phone: 800 232 7766 or 504 293 7760
Fax: 504 293 8471

**Price Index Number:**  4.0-6.3
**Price Index Standard:  virgin rubber floor
tiles, 3/16 in. thick**

### TILE AND SHEET FLOORING

SAF-DEK made from 75% post-consumer recycled rubber and 25% EPDM topcoat poured into place and handtroweled into a seamless, porous, resilient, slip-resistant surface for exterior decking, playgrounds, and other active areas. Installations include Universal Studios, DisneyWorld, and Busch Gardens in Florida.

## 408 Oscoda Plastics, Inc.

5585 N. Huron Ave., PO Box 189,
Oscoda, MI 48750
Phone: 800 544 9538   Fax: 517 739 1494

**Price Index Number:**  0.50
**Price Index Standard:  virgin rubber floor**

### TILE AND SHEET FLOORING

PROTECT-ALL commercial-grade PVC flooring from post-industrial automobile and roofing vinyl, for interior heavy-use areas. Available in sheets 1/8 in. or 1/4 in. thick, or tiles. Tiles are square cut, 18 in. by 18 in., or interlocking tongue and groove, 24 in. by 24 in. Product can be sealed with heat or chemicals. Colors available are light gray, dark gray, burgundy, blue, tan, and green. Applications include hockey rinks and community center or church multipurpose rooms. Company recycles fabric-backed vinyls through a patented process to incorporate polyester and rayon fibers for extra durability.

## 409 Plastipro Canada, Inc.

6855 Blvd. Couture, St.-Leonard, QC,
Canada H1P 3M6
Phone: 514 321 0309   Fax: 514 321 2287

**Price Index Number:**  0.90
**Price Index Standard:  virgin rubber floor**

### TILE FLOORING

PVC cushion floor tiles manufactured from recycled materials. Tile comes 12 in. by 12 in., 16 in. by 16 in., or 48 in. by 48 in. by 1/2 in. or 3/8 in. thick as square cut or interlocking with nonskid, perforated or solid surface. Commercial, residential, and industrial applications.

## 410 RB Rubber Products, Inc.

904 E. 10th, McMinnville, OR 97128
Phone: 800 525 5530 or 503 472 4691
Fax: 503 434 4455

**Price Index Number:**  0.25-0.33
**Price Index Standard:  virgin rubber matting**

### MATS AND UNDERLAYMENT

Rubber matting from 100% recycled waste tire rubber. Excellent shock absorption, resilient, soft, and durable. Used in entryways, gyms, horse stalls, and as industrial anti-fatigue carpet or underlayment. Available in 1/2-in., 3/8-in., and 3/4-in. thicknesses up to 8 ft. by 8 ft. in size. Fusing agent available for making larger mats. Charcoal color. Has rubber odor.

## 411 RCM International

978 Hermitage Rd. N.E., Rome, GA 30161
Phone: 800 328 9203　Fax: 306 295 4114

**Price Index Number:** **3.2**
**Price Index Standard:** **virgin rubber floor tiles, 3/16 in. thick**

### TILE AND SHEET FLOORING

LOCK TILE tiles made from 100% recycled PVC. Tongue-and-groove interlocking tile with raised disk surface, 19½ in. by 19½ in. by 5/16 in., provides monolithic, stable loose-laid floor that can be installed with unskilled labor. No adhesives required. Available in black and gray. TIRE TILE tiles made from recycled tires, 12 in. by 12 in. with a nonskid, nondirectional pattern. Available in a muted blackish/brown color blend. Company also manufactures 12-in. by 25-ft. runner strips made from recycled tires. Runner strips are joined together with an adhesive to achieve desired width. Typical application includes ice-skating rink walking surfaces. All products manufactured in United States.

## 412 Turtle Plastics

2366 Woodhill Rd., Cleveland, OH 44106
Phone: 216 791 2100　Fax: 216 791 7117

**Price Index Number:** **1.0**
**Price Index Standard:** **exterior vinyl surfacing**

### TILE AND SHEET FLOORING

TURTLE TILES from 100% recycled PVC in an interlocking grid surface. Tiles are 12 in. by 12 in. by ¾ in. and are available with surfaces that are either solid or perforated with triangular holes for drainage. Replacement warranty for 5 years. GRID TOP incorporates carbide grit into TURTLE TILES for wet surfaces requiring traction. TUFT STUFF tiles, 12 in. by 12 in. by ½ in. or ¾ in. thick, incorporate recycled PVC with recycled interior/exterior carpeting. Available in 13 vinyl colors and 9 carpet colors. All products designed to resist solvents, chemicals, and ultraviolet light.

## 413 W. F. Taylor Co., Inc.

11545 Pacific Ave., Fontana, CA 92337
Phone: 800 397 4583 or 800 272 4583 (tech)
Fax: 909 360 1177

**Price Index Number:** **1.10**
**Price Index Standard:** **solvent-based adhesives**

### FLOOR ADHESIVES

ENVIROTEC HEALTHGUARD provides a full line of zero calculated VOC solvent-free adhesives for carpet and vinyl flooring. Recommended by EPA.

## 414 Allegro Rug Weaving

802 S. Sherman St., Longmont, CO 80501
Phone: 800 783 1784 or 303 651 0555
Fax: 303 651 3555

**Price Index Number:** **3.0-6.0**
**Price Index Standard:** **standard nylon area rug**

### AREA RUGS

Area rugs handwoven from 8/6 Irish linen warp and 6-ply 100% wool weft using natural dyes in various colors and patterns. Company will produce custom designs. Also provides custom color dye services and will match colors to existing fabric samples. Maximum unseamed width is 12 ft. Seams are handstitched and reversible. Rugs are durable and retain their shape well and resist staining and burning. No toxic chemicals or heavy metals present in these rugs or used in their manufacture.

## 415 Bloomsburg Carpet Industries, Inc.

919 3rd Ave., New York, NY 10022
Phone: 800 336 5582 or 212 688 7447
Fax: 212 688 9218

**Price Index Number:** 2.3-3.0
**Price Index Standard:** average nylon carpet

**CARPET**
Specializing in the manufacture of woven-wool carpets with no secondary backing, up to 12 ft. wide without seams. Woven-wool carpets are noted for being stronger and more stable than modern tufted carpets. This manufacturer is one of the last woven-wool carpet mills remaining in the United States. Woven-wool carpets are typically a traditional European product.

## 416 Blue Ridge Carpet Mills

PO Box 507, 100 Progress Rd., Ellijay, GA 30540
Phone: 800 241 5945 or 706 276 2001
Fax: 706 276 2005

**Price Index Number:** 1.0
**Price Index Standard:** carpet from virgin materials

**CARPET**
Manufacturer of carpets using 100% recycled nylon from BASF. Lines include MIRADA-III, MIRADA-IV, EXCALIBUR, and NETWORK. All commercial-grade products. BASF has a program for reclaiming these products for recycling.

## 417 Carousel Carpet Mills

1 Carousel Ln., Ukiah, CA 95482
Phone: 707 485 0333   Fax: 707 485 5911

**Price Index Number:** 6.0-7.7
**Price Index Standard:** average nylon carpet

**CARPET**
Carpets made from all-natural materials: wool, cotton, and linen. Small custom mill manufactures both woven and machine-made high-end products. Will apply natural latex as backing to machine-made carpets upon request.

## 418 Chicago Adhesive

4658 W. 60th St., Chicago, IL 60629
Phone: 800 621 0220 or 773 581 1300
Fax: 773 581 2629

**Price Index Number:** 1.10
**Price Index Standard:** solvent-based adhesives

**CARPET ADHESIVES**
SAFE-SET low-toxic commercial carpet adhesives contain no ethylene glycol or solvents, although ventilation is still recommended during and immediately after installation.

## 419 Colin Campbell Ltd.

1428 W. 7th Ave., Vancouver, BC, Canada V6H ICI
Phone: 604 734 2758   Fax: 604 734 1512

**Price Index Number:** 2.8-4.0
**Price Index Standard:** average nylon carpet

**CARPET**
NATURE'S CARPET has one of the lowest VOC ratings in the industry. Manufactured from 100% natural New Zealand wool with no moth treatment, no synthetic latexes, and no chemical dyes. Both primary and secondary backings are made of jute. Distributors claim NATURE'S CARPET has worked extremely well for their clients who have chemical hypersensitivities. Carpet is biodegradable. Manufactured in Denmark.

## 420 Dura Undercushions, Inc.

8525 Delmeade Rd., Montreal, QC,
Canada H4T 1M1
Phone: 514 737 6561    Fax: 514 342 7940

**Price Index Number:** 1.0-1.2
**Price Index Standard:** virgin rubber
underlayment

### CARPET UNDERLAYMENT

DURA UNDERCUSHIONS commercial carpet padding made from ground rubber tires bonded with latex and backed with a fiberglass and cellulose mesh. Contains 92% recycled tire rubber. Open-cell structure provides excellent sound absorption and allows drying to resist mildew.

## 421 Fairmont Corp.

9759 Distribution Ave., San Diego, CA 92121
Phone: 800 621 6907 (IL) or 800 624 9616 (CA)
Fax: 312 376 3037 (IL) or 619 566 8756 (CA)

**Price Index Number:** 0.67, residential/
1.2, commercial
**Price Index Standard:** nylon/virgin rubber
underlayment

### CARPET UNDERLAYMENT

Made from ground recycled bonded polyurethane fiber. Designed for double glue-down installation in commercial applications, 10 lb. to 14 lb. Manufacturer recommends 6 lb. to 8 lb., ½ in. thick, for residential applications. Gensler and Associates has installed DUBL-BAC product in their offices in San Francisco, Calif.

## 422 Fibreworks

1729 Research Dr., Louisville, KY 40299
Phone: 800 843 0063 or 502 499 9944
Fax: 800 843 0063 or 502 499 9880

**Price Index Number:** 0.80, floor/1.4, wall
**Price Index Standard:** average nylon carpet/
cotton fabric

### CARPET

Sisal and jute floor and wall coverings without backing in widths of 4 ft. to 12 ft., up to 100 ft. in length; 20 colors available.

## 423 Flokati Wool Rugs

The Boulevard, 2373 Broadway, #930,
New York, NY 10024
Phone: 888 356 5284 or 212 875 0877
Fax: 212 875 0977

**Price Index Number:** 2.0-2.8
**Price Index Standard:** average nylon area rug

### AREA RUGS

Flokati wool area rugs imported from Greece. These are long-stranded rugs made from 100% New Zealand Drysdale wool.

## 424 Hendricksen Naturlich

7129 Keating Ave., Sebastopol, CA 95472-3805
Phone: 707 829 3959    Fax: 707 829 1774

**Price Index Number:** 1.7-6.7
**Price Index Standard:** average nylon carpet

### CARPET

Natural-wool carpets in many patterns, textures, and colors. Carpet underlayment from natural-fiber post-industrial remnants, ½ in. thick, 8 lb. Company also manufactures sisal, jute, and seagrass rugs and carpets available in many textures, colors, and patterns.

## 425 Hendricksen Naturlich

6761 Sebastopol Ave., Suite 7,
Sebastopol, CA 95472-3805
Phone: 707 829 3959   Fax: 707 829 1774

**Price Index Number:** 1.25/2.20, wool and camel hair
**Price Index Standard:** nylon underlayment, ½ in.

**CARPET UNDERLAYMENT**
From post-industrial natural-fiber remnants heat bonded together to produce an underlayment without toxic outgassing. Comes in ⁵⁄₁₆-in., ⅜-in., and ½-in. thick, 8-lb. pad. For an extra soft feel, a ⁷⁄₁₆-in. wool and camel hair underlayment is also available.

## 426 Homasote Co.

PO Box 7240, W. Trenton, NJ 00620-0240
Phone: 800 257 9491 or 609 883 3300
Fax: 609 883-3300

**Price Index Number:** 1.5
**Price Index Standard:** nylon underlayment, ½ in.

**CARPET UNDERLAYMENT**
440 CARPETBOARD and COMFORTBASE are compressed paper boards from 100% recycled newsprint cellulose. Both are available as 4-ft. by 8-ft. by ½-in. sheets, and the COMFORTBASE is also available as a 4-ft. by 4-ft. by ½-in. sheet with scoring to conform to irregular surfaces. Treated to protect against termites, rot, and fungi. No urea formaldehyde or asbestos.

## 427 Image Carpets, Inc.

PO Box 5555, Armuchee, GA 30105
Phone: 800 722 2504 or 800 241 7597
Fax: 706 235 8584

**Price Index Number:** 0.3-1.0
**Price Index Standard:** average nylon carpet

**CARPET**
DURATRON, PRESERVATION, and WEARLON carpets are manufactured with plastic fibers from recycled PET. WEARLON is made from 100% recycled coke bottle-grade PET plastics, which provide a more durable fiber than virgin material. PET fibers are naturally stain resistant, not requiring chemical treatment like other nylon carpets. Carpet is ¹⁄₁₀ gauge, 25 oz. to 75 oz. Machine-made with action-backed latex in 15 ft. widths.

## 428 Langhorne Carpet Co.

PO Box 7175, 201 W. Lincoln Highway
Penndel, PA 19047-0824
Phone: 215 757 5155   Fax: 215 757 2212

**Price Index Number:** 5.0, custom/ 4.0, stock
**Price Index Standard:** average nylon carpet

**CARPET**
Manufacturer of custom carpets from 100% wool face yarns and backed with cotton and jute yarns. Specializing in duplicating carpet designs and colors from the late 19th or early 20th centuries, the company makes carpets on looms similar to 19th-century originals. Carpets are woven in 27-in. widths and sewn together in the traditional method. Minimum of 30 linear yards for custom orders.

## 429 Merida Meridian, Inc.

643 Summer, Boston, MA 02210
Phone: 800 345 2200   Fax: 617 268 9594

**Price Index Number:** 0.80-2.0
**Price Index Standard:** average nylon carpet

**CARPET**
Sisal, sisal plus wool blend, coir, jute, and seagrass broadloom and area rugs with latex backing. Maximum unseamed width is 13 ft. Newest product is recycled paper plus sisal combination carpet, with latex backing. Made in Belgium.

## 430 Reliance Carpet Cushion

137 E. Alondra St., Gardena, CA 90248
Phone: 800 522 5252 or 213 321 2300
Fax: 310 323 4018

**Price Index Number:** 0.85-1.0
**Price Index Standard:** new nylon
underlayment, ½ in.

**CARPET UNDERLAYMENT**
Carpet-fiber underlayment manufactured from textile waste and recycled carpet fibers.

## 431 Savnik and Co. Tailors in Wool

601 McClary Ave., Oakland, CA 94621
Phone: 800 872 8645 or 510 568 4628
Fax: 510 568 3150

**Price Index Number:** 3.0
**Price Index Standard:** residential and
commercial rugs

**CARPET**
Custom manufacturer of 100% wool area rugs and wall-to-wall carpets.

## 432 Sutherlin Carpet Mills

342 Westway, Orange, CA 92665
Phone: 714 447 0792   Fax: 714 449 1253

**Price Index Number:** 1.0
**Price Index Standard:** nylon underlayment

**CARPET UNDERLAYMENT**
Company has developed product from recycled nylon for chemically hypersensitive people. Willing to work closely with customers to meet specific needs. Minimum order of 50 yards.

## 433 Sutherlin Carpet Mills

342 Westway, Orange, CA 92665
Phone: 714 447 0792   Fax: 714 449 1253

**Price Index Number:** 1.3-1.7
**Price Index Standard:** average nylon carpet

**CARPET**
Natural-fiber machine-made carpets. Company also has developed special products for chemically hypersensitive people including SAFGUARD carpet from virgin untreated nylon using low-toxic dyes, latex primary backing, urethane secondary backing, and carpet underlayment from recycled nylon. Lowest in toxins is HYPERSENSITIVE carpet, which manufacturer claims to be 97% toxin free with a nonlatex backing. Minimum order of 50 yards. Willing to work closely with customers to meet specific needs.

## 434 Talisman Mills, Inc.

6000 W. Executive Dr., Suite H,
Mequon, WI 53092
Phone: 800 482 5466 or 414 242 6183
Fax: 414 242 6751

**Price Index Number:** 1.0-1.8
**Price Index Standard:** average nylon carpet

**CARPET**
ENVIRELON from post-consumer 100% recycled PET plastic. Commercial quality. PET plastics provide a more durable fiber than virgin material. PET fibers are naturally stain resistant, not requiring chemical treatment like other nylon carpets. No representative in California.

## 435 Terra Nativa Sisal

PO Box 313, Excelsior, MN 55331
Phone: 800 287 3144 or 612 933 7773
Fax: 612 933 7374

**Price Index Number:**   **1.0, BROADLOOM**
                      **with backing**
**Price Index Standard:**  **average nylon**
                        **area rug**

### AREA RUGS
Sisal broadloom and area rugs made from 100% sustainably harvested sisal fiber from Brazil. Maximum broadloom width without seam is 12 ft.

---

## SPECIAL FLOORING                                                                  09700

## 436 Exerflex

6801 Lake Plaza, Suite A-105,
Indianapolis, IN 46220
Phone: 800 428 5306   Fax: 317 842 5384

**Price Index Number:**   **1.0+**
**Price Index Standard:**  **wood floor**

### PLASTIC FLOORING
ECO-DESIGN plastic planks manufactured from all-recycled plastic are an alternative to wood flooring. Planks are solid, ⅞ in. thick with wood-grain finish available in 6 colors.

---

## 437 Multi-Tech Ltd.

2318 3rd St., Sioux City, IA 51101
Phone: 800 452 4374 or 712 258 5975
Fax: 712 233 2879

**Price Index Number:**   **1.0-1.5**
**Price Index Standard:**  **acrylic flooring**

### TILE AND SHEET FLOORING
DURA-GUARD 9000B is a flooring material manufactured from scrap PVC. Suitable as a resilient surface for heavy-duty wear areas including tennis courts and pool decks. Available in blue, burgundy, green, tan, light gray, and dark gray with a smooth or nonskid surface. Available sizes are 30 in. by 5 ft., 5 ft. by 5 ft., 5 ft. by 8 ft., and also in 24-in. by 24-in. and 12-in. by 12-in. tiles. Comes with 2-year limited warranty.

---

## SPECIAL COATINGS                                                                  09800

## 438 BathCrest, Inc.

2425 S. Progress Dr., Salt Lake City, UT 84119
Phone: 801 972 1110 or 800 826 6790
Fax: 801 977 0328

**Price Index Number:**   **0.12**
**Price Index Standard:**  **standard tub**
                        **replacement**

### PORCELAIN SURFACE
GLAZECOTE is covalent-bonded, porcelain-like coating that is chemically bonded to hard inorganic surfaces for a hard, glossy finish. Typically used to repair finishes on bathtubs, sinks, and tile. Available in colors matching those of most major bathroom fixture manufacturers. Product is applied by a factory representative and comes with a 5-year warranty. This process offers a less expensive alternative to replacing fixtures due to surface wear or color changes. Manufacturer claims the finished product to be nontoxic.

## 439 Key Solutions

PO Box 5090, Scottsdale, AZ 85261
Phone: 800 776 9765 or 602 948 5150
Fax: 602 998 7405

**Price Index Number:** 0.75, per gallon cost comparison
**Price Index Standard:** Benjamin Moore top latex

### CERAMIC INSULATING PAINT

THERMOFLEX 3 insulating paint is an acrylic latex paint with ceramics added designed to reduce heat gain in exterior applications and heat loss in interior applications. Coverage is 80 to 100 sq. ft. per gal. THERMOFLEX 3 used on Arizona roofs to reduce heat gain.

## 440 AFM Enterprises, Inc.

350 W. Ash St., Suite 700, San Diego, CA 92101
Phone: 619 239 0321    Fax: 619 239 0565

**Price Index Number:** 1.0
**Price Index Standard:** Benjamin Moore top latex

### PAINTS AND STAINS

SAFECOAT paints and stains. Manufacturer of a complete line of nontoxic paints, sealers, cleaners, adhesives, stains, and finishes for both residential and commercial use. Water-based products without formaldehyde, fungicide, and mildewcide. Meet strictest VOC environmental regulations. Products developed specifically for chemically sensitive people in consultation with environmental medicine physicians.

## 441 Auro-Sinan Co.

PO Box 857, Davis, CA 95617
Phone: 916 753 3104    Fax: 916 753 3104

**Price Index Number:** 2.0 standard/ 2.35 professional
**Price Index Standard:** Benjamin Moore top latex

### PAINTS

AURO, a full line of natural-based paints in a standard and professional line. No petroleums or plastic ingredients. Solvents are less than 1%, no biocides, fungicides, lead, mercury or crystalline silica in paint. Product performs like a typical top-quality latex paint at temperatures greater than 50°F with about 10% less coverage. Has a 1-year shelf life. Product has been in use for 12 years with good results. Company broke away from Livos to establish stricter health and environmental guidelines.

## 442 Benjamin Moore and Co.

51 Chestnut Ridge Rd., Montvale, NJ 07645
Phone: 800 826 2623    Fax: 201 573 9600

**Price Index Number:** 1.2
**Price Index Standard:** standard top-quality latex

### PAINTS

PRISTINE is an acrylic latex paint that the manufacturer claims is without VOCs. Available as an interior primer and sealer in flat, eggshell, and semigloss.

## 443 Bollen International, Inc.

16479 Dallas Parkway, Suite 500, Dallas, TX 75248
Phone: 800 248 4808 or 214 250 2964
Fax: 214 250 1370

**Price Index Number:** 3.0
**Price Index Standard:** Benjamin Moore top latex

### PAINTS

FERROXTON-W is an exterior water-based paint with a low VOC content. Also nonflammable and odorless, it is available in metallic coating with textures that look like stainless steel, copper, brass, and blue metal. CRAFTON and CRAFTON PLUS are multicolored, water-based emulsion paints that create a fabric-like texture on a surface. All products are sprayed on.

## 444 California Products Corp.

169 Waverly St., Cambridge, MA 02139-0569
Phone: 800 533 5788   Fax: 617 547 6934

**Price Index Number:** **2.0**
**Price Index Standard:** **Benjamin Moore
top latex**

**PAINTS**
AQUAFLECK is a residential, interior use, spray-on multicolor water-based undercoat and finish coat of 100% acrylic latex paint. It is nontoxic, free of solvents, and has a Class A fire rating. This product can be used to replace vinyl wall coverings or solvent-based multicolor paints.

## 445 Coronado Paint Co.

308 Old County Rd., Edgewater, FL 32132
Phone: 800 883 4193   Fax: 800 394 9022

**Price Index Number:** **0.80**
**Price Index Standard:** **Benjamin Moore
top latex**

**PAINTS**
SUPREME COLLECTION is a nontoxic, no VOC latex paint, which comes in flat (914-1), eggshell (916-1), or semigloss (926-1). Color availability is limited to tintable whites.

## 446 Eco Design/Natural Choice

1365 Rufina Circle, Santa Fe, NM 87505
Phone: 800 621 2591 or 505 438 3448
Fax: 505 438 0199

**Price Index Number:** **1.6/2.0/1.0**
**Price Index Standard:** **Benjamin Moore
top latex**

**PAINTS AND STAINS**
LIVOS DUBRON, BIOSHIELD, and CASEIN paints and wood finishes are reported to be appropriate for allergy-prone and chemically sensitive people. Some products in the LIVOS line contain a biodegradable solvent, and the DUBRAN contains no isoalaphatic to suppress the citrus odor. CASEIN is the least reactive product for allergy-prone individuals. LIVOS is made in Germany and has been in use in United States for 10 years. BIOSHIELD line is made from all-natural materials.

## 447 Glidden Paint and Wallcovering

1900 N. Jose Ln., Carrolington, TX 75006
Phone: 800 553 1420 or 800 221 4100
Fax: 972 417 7501

**Price Index Number:** **1.5**
**Price Index Standard:** **latex paint
with VOCs**

**PAINTS**
SPRED 2000 is a latex paint without VOCs. It is available in a variety of whites that can be custom tinted to produce a light pastel color. Manufacturer claims this product to be nontoxic.

## 448 The Green Paint Co.

PO Box 430, 9 Main St., Manchaug, MA 01526
Phone: 800 477 1992 or 508 234 5777
Fax: 508 476 1201

**Price Index Number:** **0.60**
**Price Index Standard:** **virgin latex paint**

**PAINTS**
Paint manufactured from recycled paint that is 90% post-consumer waste. Product meets or exceeds standards for virgin paint. Many paint types are available, including exterior latex house paint, interior latex eggshell finish, interior latex flat finish, exterior oil-based primer, solid exterior oil stain, and urethane reinforced alkyd floor enamel. Packaging is recycled and recyclable. Samples are available. This product is available from distributors in New England or directly from the company. Recycled paint comes from the hazardous waste collection program. White and earth tones are available without a minimum order; any other colors require a 500-gal. minimum.

## 449 Kelley-Moore Paint Company

5101 Raley Blvd., Sacramento, CA 95838
Phone: 800 874 4436 (San Carlos) or 916 921 0165
Fax: 916 921 0184

**Price Index Number:** 0.5-0.7
**Price Index Standard:** virgin latex paint

**PAINTS**
The company accepts used latex paint at its San Carlos, Calif., headquarters for reprocessing into medium-grade exterior paint for sale at the same location. Available in a limited assortment of colors.

## 450 Miller Paint Co.

317 S.E. Grand Ave., Porland, OR 97214
Phone: 800 852 3254 or 503 233 4491
Fax: 503 238 6289

**Price Index Number:** 1.0
**Price Index Standard:** Benjamin Moore top latex

**PAINTS**
Solvent-free, low-biocide, and no fungicide paint product line. The paint is also free of mercury, lead, and crystalline silica. Product performs like a typical top-quality latex paint. This is a small company that makes fresh paint on order, thereby avoiding the need for biocides for storage. Product has been in use for 8 years with good results.

## 451 Murco Wall Products, Inc.

300 N.E. 21st St., Fort Worth, TX 76106
Phone: 800 446 7124 or 817 626 1987
Fax: 817 626 0821

**Price Index Number:** 0.60
**Price Index Standard:** Benjamin Moore top latex

**PAINTS**
LE (gloss) or GF (flat) MURCO is a low-odor, no fungicide premium-quality paint. Product performs like latex paint. In high-moisture areas, mildew growth may increase. No mercury or lead in the paint.

## 452 Old-fashioned Milk Paint Co.

PO Box 222, Groton, MA 01450
Phone: 508 448 6336   Fax: 508 448 2754

**Price Index Number:** 2.0
**Price Index Standard:** Benjamin Moore top latex

**PAINTS**
Milk-based paints are manufactured by hand from milk proteins and shipped out in powder form. Available in 16 colors that can be easily combined by the user to create custom colors. The paint's appearance is flat and somewhat streaky. It can pick up water stains and should be sealed with a water-based sealer in areas where water is liable to be splashed on a painted surface. This product is popular with people restoring old houses and antiques, and where chemical sensitivity is an issue, such as in hospitals and schools. This paint is safe for people and the environment, as it is made without fungicides or mildewcides. This is a small, family-run business that does not advertise.

## 453 Pace Chem Industries, Inc.

PO Box 1946, Santa Ynez, CA 93460
Phone: 800 350 2912 or 805 667 2140
Fax: 805 667 2145

**Price Index Number:** **1.60**
**Price Index Standard:** **Benjamin Moore top latex**

**PAINTS**
CRYSTAL SHIELD claims to be a low VOC, chemically safe, nontoxic latex paint that also acts as a sealer. Available in white and pastel tints. A 15% discount is available to architects.

## 454 Palmer Industries

10611 Old Annapolis Rd., Frederick, MD 21701
Phone: 301 898 7848　　Fax: 301 898 3312

**Price Index Number:** **0.65, #9400W/ 0.95-1.1, #8600-1**
**Price Index Standard:** **Benjamin Moore top latex**

**PAINTS**
8600-1SEAL is a modified latex clear primer sealer and vapor barrier with a Class 1 fire rating and no VOCs. It is approved by the FDA to be used where it may come into contact with foods. Available in flat and gloss. 9400 is a solvent-free concrete sealer that prevents leakage of efflorescence. 9400W is a solvent-free water repellent impregnator for wood.

## 455 Rasmussen Paint Co.

12655 S.W. Beaverdam Rd., Beaverton, OR 97005
Phone: 800 992 6692 or 503 644 9137
Fax: 503 644 9131

**Price Index Number:** **0.80-1.0**
**Price Index Standard:** **virgin latex paint**

**PAINTS**
100% and 50% recycled latex paints for interior and exterior use, collected through Portland Metro's household hazardous waste facility. Available as both a primer and a surface coating for residential or industrial applications. Purchase price includes delivery to site in Oregon and Washington.

## 456 Republic Paints

1128 N. Highland Ave., Hollywood, CA 90038
Phone: 213 957 3060　　Fax: 213 957 3061

**Price Index Number:** **0.7-1.5**
**Price Index Standard:** **Benjamin Moore top latex**

**PAINTS**
The company supplies nontoxic paints with no VOCs to Hollywood studios and sets. Also available is a clear acrylic finish for interior or exterior that comes in flat, semigloss, satin, or gloss.

## 457 Savogran

259 Lenox St., Norwood, MA 06062
Phone: 800 225 9872　　Fax: 617 762 1095

**Price Index Number:** **1.32**
**Price Index Standard:** **standard paint cleaners**

**PAINT CLEANERS**
CLEANSAFER removes both oil and latex paint from brushes. Slightly flammable according to OSHA guidelines. Available in pints by special order through Servistar (product #1191) or other hardware stores. Manufacturer claims product to be nontoxic if not ingested.

## 458 Spectra-tone Paint Co.

1595 E. San Bernadino Ave.,
San Bernardino, CA 92408
Phone: 800 272 4687   Fax: 909 478 3499

**Price Index Number:** **0.6-0.79**
**Price Index Standard:** **Benjamin Moore
top latex**

### PAINTS
NO VOC ENVIRO is a nontoxic acrylic latex paint with no VOCs. Available in flat, eggshell, and semigloss. Polyphase, made by Troy Manufacturing and considered safe by the EPA, has been added as a fungicide. Kaiser Permanente Foundation Hospitals of Northern California are now specifying this product for their patient rooms. Their staff has studied this paint and considers a patient room safe for occupancy 2 hours after painting.

## 459 Wellborn Paint

741 S. Huron, Denver, CO 80223
Phone: 800 228 0883 or 303 778 6728
Fax: 303 778 7066

**Price Index Number:** **1.0**
**Price Index Standard:** **Benjamin Moore
top latex**

### PAINTS
Manufacturer of paints with low VOCs, no lead, and no mercury. Owned by Dunn Edwards Paints. Distribution limited to Colorado, New Mexico, and El Paso, Tex.

## 460 William Zinsser and Co.

173 Belmont Dr., Somerset, NJ 08875-1285
Phone: 908 469 8100   Fax: 908 563 9774

**Price Index Number:** **1.0, BULLSEYE/
1.2, BIN**
**Price Index Standard:** **Benjamin Moore
top latex**

### PAINTS AND SHELLACS
BULLSEYE 1-2-3 PRIMER-SEALER is a water-based interior-exterior primer. Breathable and flexible, it adheres to many substrates including wood, gypsum board, masonry, and smooth surfaces such as glass or tile. It is odorless and cleans with soap and water. Available in many hardware stores. BIN is a natural shellac harvested in Asia from used cocoons dissolved in ethanol alcohol. This alcohol is the only main VOC ingredient.

# WALL COVERINGS                                    09950

## 461 Crown Corp. NA

3012 Huron St., Suite 101, Denver, CO 80202
Phone: 800 422 2099 or 303 292 1313
Fax: 303 292 1933

**Price Index Number:** **1.0**
**Price Index Standard:** **vinyl wallpaper**

### WALLPAPER
Distributor of traditional anaglypta supadurable wall coverings made from 90% recycled cotton and 10% wood from sustainably harvested forests in Finland. The product is nontoxic and doesn't have chemical additives or glues. No backing material. Made in England.

## 462 Design Materials, Inc.

241 S. 55th St., Kansas City, KS 66106
Phone: 800 654 6451 or 913 342 9796
Fax: 913 342 9826

**Price Index Number:** **2.3, sisal**
**Price Index Standard:** **vinyl wallpaper**

### WALL AND FLOOR COVERINGS
Distributor of wall and floor coverings manufactured in Mexico using natural ingredients of fiber sisal, jute, and wool. Floor products include both carpet and area rugs.

### 463 Flexi-Wall Systems

PO Box 89, Liberty, SC 29657
Phone: 800 843 5394 or 803 843 3104
Fax: 803 843 9318

**Price Index Number:** **0.6, undersurface/**
**1.0-1.30, finished**
**Price Index Standard:** **vinyl wallpaper**

**PLASTER WALL FABRIC**
FASTER PLASTER is a low-toxicity wall fabric with a jute substrate supporting an uncrystallized plaster coating. Fabric is applied to wall with a water-based adhesive, causing crystallization of the plaster coating. Available as a finished wall surface or as an undersurface for other applications, such as wallpaper. Class A fire rating and zero flame spread.

### 464 KORQ, Inc.

400 E. 56th St., New York, NY 10022
Phone: 212 758 2593   Fax: 212 758 0025

**Price Index Number:** **2.0, cork wallpaper**
**Price Index Standard:** **standard cotton**
**wallcovering**

**WALLPAPER**
Cork wallpaper, fabrics, and shower tiles. Wallpaper and fabric has suede texture. Tiles come 12 in. by 12 in., 16 in. by 16 in., and 30cm by 90cm, each ¼ in. thick. Available in 11 colors and natural. Naturally anti-bacterial. Cork is a renewable resource harvested on a 9-year cycle. Company is the export office for Sardinian cork producers.

### 465 Pallas Textiles

1330 Bellevue St., Green Bay, WI 54302
Phone: 800 454 9796 x2660 or 414 468 2660
Fax: 414 468 2661

**Price Index Number:** **2.0**
**Price Index Standard:** **vinyl wallpaper**

**WALL COVERINGS**
EARTH PAPER is a wall covering manufactured from wood pulp, stone powder, and straw. Available in a variety of subtle, textured earth tones, such as sand, bark, clay, adobe, and greige. Rolls are 36½ in. wide by 50 yd. long.

## TOILET PARTITIONS 10170

### 466 Santana Laminations

PO Box 2021, Scranton, PA 18501
Phone: 800 233 4701 or 717 343 7921
Fax: 717 348 2959

**Price Index Number:** **1.0**
**Price Index Standard:** **virgin materials**

**PLASTIC TOILET PARTITIONS**
Manufacturer of recycled plastic toilet and shower compartments, vanity tops, and locker-room benches. Company claims the products contain a minimum of 10% recycled content.

## FIREPLACES AND STOVES 10300

### 467 Biofire, Inc.

3220 Melbourne, Salt Lake City, UT 84106
Phone: 801 486 0266   Fax: 801 486 8100

**Price Index Number:** **4.0-8.0**
**Price Index Standard:** **standard masonry**
**fireplace**
**(0% efficiency)**

**MASONRY FIREPLACES**
BIOFIRE fireplace is a masonry wood-burning fireplace with a TUV-tested efficiency of 90%. This product is manufactured in Austria by SUPERFIRE and is custom designed for the client's space. Available with a large selection of tiles and accessories. Many of these fireplaces are especially beautiful.

## 468 Envirotech

PO Box 323, Vashon Island, WA 98070
Phone: 800 325 3629 or 206 463 3722
Fax: 206 463 6335

**Price Index Number:**   **2.8-3.8**
**Price Index Standard:**   **standard masonry**
                                  **fireplace**
                                  **(0% efficiency)**

**MASONRY FIREPLACES**
ENVIROTECH radiant fireplace kit is a masonry heater in modular kit form. Made of castable refractory firebrick. Pieces are tongue and groove and numbered for ease of installation. High efficiency, low emissions, and creosote free.

## 469 Firespaces

921 S.W. Morrison, Suite 440, Portland, OR 97205
Phone: 503 227 0547    Fax: 503 227 0548

**Price Index Number:**   **1.8**
**Price Index Standard:**   **standard masonry**
                                  **fireplace**
                                  **(0% efficiency)**

**MASONRY FIREPLACES**
MOBURG MRC fireplace is a masonry wood-burning fireplace. It is a closed combustion system with glass doors and outside air intakes. With a 54% efficiency, this product has been approved as a clean-burning fireplace by Colorado, Washington, and parts of California. Fireplace is available in 36-in. and 42-in. fire view widths with rectangular or arched openings. Thermodynamics follows a modified Rumford design with Rosen design features. Exhaust gas burns off creosote at 2000°F and passes through channels to transfer heat to living space.

## 470 Jotul USA, Inc.

PO Box 1157, Portland, ME 04104
Phone: 207 797 5912    Fax: 207 772 0523

**Price Index Number:**   **0.85-2.2**
**Price Index Standard:**   **standard residential**
                                  **furnace**

**WOOD AND GAS STOVES**
Cast-iron wood and gas stoves for space heating. Manufactured from 100% recycled steel with enamel finish utilizing recycled lead.

## 471 Kent Valley Masonry

23631 S.E. 216 St., Maple Valley, WA 98038
Phone: 206 432 0134    Fax: 206 413 1771

**Price Index Number:**   **2.7-6.8**
**Price Index Standard:**   **standard masonry**
                                  **fireplace**
                                  **(0% efficiency)**

**MASONRY FIREPLACES**
Supplier and builder of GRUNDOFEN, HEAT-KIT, and TULIKIVI masonry heaters. Custom-made tile, fieldstone, brick, and soapstone wood heaters beautifully designed and highly efficient. Available with bake oven option. Main area of business is the Pacific Northwest.

## 472 Lopez Quarries

111 Barbara Ln., Everett, WA 98203
Phone: 206 353 8963    Fax: 206 742 3361

**Price Index Number:**   **4.0-8.0**
**Price Index Standard:**   **standard masonry**
                                  **fireplace**
                                  **(0% efficiency)**

**MASONRY FIREPLACES**
Distributor and builder of a variety of high-efficiency masonry wood-burning fireplaces including FRISCH-ROSIN, BARTSCH, CRONSPESEN, BIOFIRE, and TULIKIVI.

### 473 Temp-Cast Enviroheat Ltd.

PO Box 94059, 3332 Young St.,
Toronto, ON, Canada M4N 3R1
Phone: 800 561 8594 or 416 322 6084
Fax: 416 486 3624

**Price Index Number:** 1.4
**Price Index Standard:** standard masonry
fireplace
(0% efficiency)

**MASONRY FIREPLACES**

TEMP-CAST is a masonry modular wood-burning or natural or propane gas heater. Wood-burning model weighs 2,800 lb. with an EPA-rated heating efficiency of 65% and a combustion efficiency of 94%. Emissions are 1.3 gm/kg as determined by an independent testing lab. It is approved for use in most regions of the United States. TEMP-CAST heaters come in kit form to be assembled on-site from cast refractory masonry blocks made from 90% recycled materials. The kit provides the cove only, which is custom finished by a mason typically using brick, stone, or tile. Wood-burning heater is available with bake oven, corner, or see-through designs. Natural gas heaters are rated for 30,000 BTUH input, and propane heaters are rated for 28,000 BTUH input. The gas heater is a thermostatically controlled direct vent design with a sealed firebox and an air-intake and vent system.

### 474 Vermont Castings, Inc.

Rt. 10, Box 501, Bethel, VT 05032
Phone: 800 227 8683 or 802 234 2300
Fax: 802 234 2341

**Price Index Number:** 1.1-2.4, wood stove
**Price Index Standard:** standard residential
gas furnace

**FIREPLACES AND STOVES**

Handsome cast-iron wood and gas stoves, fireplaces, fireplace inserts, and pellet stoves. Catalytic, highly efficient, and clean burning. Hearth accessories available. Products are manufactured from 100% post-consumer iron from radiators and brake drums.

### 475 Humane Manufacturing Co.

805 Moore St., Baraboo, WI 53913-2796
Phone: 800 369 6263 or 608 356 8336
Fax: 608 356 8338

**Price Index Number:** 3.1
**Price Index Standard:** vinyl runner

**RUBBER WALK**

ROOF-GARD rubber walk cover to protect roofs from foot traffic. Manufactured from recycled rubber tires with no filler added. Available in the following sizes: 4 ft. by 6 ft., 2 ft. by 6 ft., or 3 ft. by 4 ft. in thicknesses of ⅜ in., ½ in., or ¾ in. with diamond or button patterns. LOKTUFF and SOFTUFF are interlocking flooring systems available in various colors. SOFTUFF has a softer finish than LOKTUFF. Manufactured from 95% recycled rubber.

### 476 Structural Plastics

2750 Lippincott Blvd., Flint, MI 48507
Phone: 800 523 6899 or 810 743 2799
Fax: 810 743 2799

**Price Index Number:** 1.0
**Price Index Standard:** wood shelves

**PLASTIC SHELVING**

Manufacturer of DURASHELF recycled plastic utility storage shelving. Claims 100% recycled content.

### 477 Zwiers, Don and Associates

PO Box 2278, Joliet, IL 60434
Phone: 800 438 7395 or 815 729 3326
Fax: 815 741 0058

**Price Index Number:** **0.90**
**Price Index Standard:** **steel frame with particleboard shelves**

**FIBERGLASS SHELVING**
Manufacturer of recycled fiberglass, heavy-duty utility shelving, 18 in. or 24 in. deep. Claims 100% recycled content.

---

## LOADING DOCK EQUIPMENT                                                          11160

### 478 Durable Corp.

75 N. Pleasant St., Norwalk, OH 44857
Phone: 800 537 1603   Fax: 800 537 6287

**Price Index Number:** **0.75, rubber speed bump**
**Price Index Standard:** **asphalt speed bump**

**DOCK BUMPER**
Dock bumpers, removable speed bumps, and floor mats made from recycled truck tires.

---

## WASTE CHUTES AND COLLECTORS                                                          11175

### 479 Feeny Manufacturing Co.

PO Box 191, Muncie, IN 47308
Phone: 317 288 8730   Fax: 317 288 0851

**Price Index Number:** **1.4**
**Price Index Standard:** **standard corner-cabinet lazy Susan**

**WASTE MANAGEMENT SYSTEM**
SPIN-A-BIN kitchen recycling organizer replaces the standard corner-cabinet lazy Susan with three 32-quart polyethylene bins suspended from a rotating steel frame. Each bin can be readily removed for emptying.

---

## WATER SUPPLY AND TREATMENT EQUIPMENT                                                          11200

### 480 Aquazone Products Co.

79 Bond St., Elk Grove Village, IL 60007
Phone: 847 439 4454   Fax: 847 439 4033

**Price Index Number:** **3.0**
**Price Index Standard:** **chlorine disinfection, whole house**

**OZONE PURIFICATION SYSTEM**
Complete line of UV-type ozone generation equipment for purification of potable water and also for swimming pools, spas, and hot tubs. Venturi and compressor-driven models. High-quality equipment.

---

### 481 Clearwater Tech.

PO Box 15330, San Luis Obispo, CA 93406
Phone: 800 262 0203 or 805 549 9724
Fax: 805 549 0306

**Price Index Number:** **2.5-3.0**
**Price Index Standard:** **chlorine disinfection, whole house**

**OZONE PURIFICATION SYSTEM**
Manufacturer of both corona-discharge and UV-type ozone generation equipment for the purification of potable water.

## 482 Del Industries

3428 Bullock Ln., San Luis Obispo, CA 93401
Phone: 800 676 1335 or 805 541 1601
Fax: 805 541 8459

**Price Index Number:** 2.5-3.0
**Price Index Standard:** chlorine disinfection, whole house

**OZONE PURIFICATION SYSTEM**
Manufacturer of UV-type ozone generation equipment for the purification of potable water.

## 483 Gas Purification System

3213 W. Hampden Ave., Englewood, CO 80110
Phone: 800 722 9106 or 303 781 9706
Fax: 303 781 9675

**Price Index Number:** 2.5-3.0
**Price Index Standard:** chlorine disinfection, whole house

**OZONE PURIFICATION SYSTEM**
Manufacturer of UV-type ozone generation equipment for the purification of potable water.

## 484 Heliotrope General

3733 Kenora Dr., Spring Valley, CA 91977
Phone: 800 552 8838 or 619 460 3930
Fax: 619 460 9211

**Price Index Number:** 2.5-3.0
**Price Index Standard:** chlorine disinfection, whole house

**OZONE PURIFICATION SYSTEM**
Manufacturer of UV-type ozone generation equipment for the purification of potable water.

## 485 Lifeguard Purification Systems

4306 W. Osborne Ave., Tampa, FL 33614
Phone: 800 678 7439 or 813 875 7777
Fax: 813 871 6250

**Price Index Number:** 2.5-3.0
**Price Index Standard:** chlorine disinfection, whole house

**OZONE PURIFICATION SYSTEM**
Manufacturer of both corona-discharge and UV-type ozone generation equipment for the purification of potable water.

## 486 Oxygen Tech.

8229 Melrose, Lenexa, KS 66214
Phone: 913 894 2828   Fax: 913 894 5455

**Price Index Number:** 2.5-3.0
**Price Index Standard:** chlorine disinfection, whole house

**OZONE PURIFICATION SYSTEM**
Manufacturer of corona-discharge-type ozone generation equipment for the purification of potable water.

## 487 Ozotech, Inc.

2401 Oberlin Rd., Yreka, CA 96097
Phone: 916 842 4189   Fax: 916 842 3238

**Price Index Number:** 2.5-3.0
**Price Index Standard:** chlorine disinfection, whole house

### OZONE PURIFICATION SYSTEM
Manufacturer of corona-discharge-type ozone generation equipment for the purification of potable water.

## 488 ReWater Systems, Inc.

438 Addison Ave., Palo Alto, CA 94301
Phone: 415 324 1307

**Price Index Number:** 1.4, irrigation system + rewater system
**Price Index Standard:** conventional subsurface irrigation system

### WATER CONSERVATION SYSTEM
Water conservation system filters water from showers, tubs, clothes washers, and bathroom and laundry sinks for use in subsurface irrigation system. Reduces residential water consumption by 25% to 45%. System storage tanks are manufactured from 25% recycled plastic.

## 489 Amana Refrigeration

2800 220th Trail, Amana, IA 52204-0001
Phone: 800 843 0304 or 319 622 5511
Fax: 319 622 2977

**Price Index Number:** 1.0
**Price Index Standard:** standard residential refrigerator

### REFRIGERATOR
Manufacturer of an efficient 20-cu.-ft., top-mounted refrigerator/freezer, model #BM20TBW, with an average energy consumption of $50 per year at 6.67¢ per kwh. Uses a non-CFC refrigerant, R-134a.

## 490 Asko, Inc.

PO Box 851805, Richardson, TX 75085-1805
Phone: 972 644 8595   Fax: 972 644 8593

**Price Index Number:** 1.25
**Price Index Standard:** standard residential dishwasher

### DISHWASHER
Manufacturer of the most efficient dishwasher, model #1475, which uses 4.6 gallons for each wash and 471 kwh per year. Made in Sweden for the U.S. market.

## 491 General Electric

Customer Relations, AP6-129, Louisville, KY 40225
Phone: 800 626 2000 or 502 452 4311
Fax: 502 452 0739

**Price Index Number:** 1.4
**Price Index Standard:** standard residential refrigerator

### REFRIGERATOR
Manufacturer of an efficient 22-cu.-ft., top-mounted refrigerator/freezer with an average U.S. energy consumption of $52 per year, model #CTH21GAS. Uses a non-CFC refrigerant.

## 492 General Electric

Customer Relations, AP6-129, Louisville, KY 40225
Phone: 800 626 2000 or 502 452 4311
Fax: 502 452 0739

**Price Index Number:** 1.5-2.3
**Price Index Standard:** standard residential dishwasher

### DISHWASHER
Manufactured by Bosch for GE, an efficient dishwasher, Monogram ZBD4300SWH, uses 7 gallons of water for each wash. See Robert Bosch Corp. (#495) for more information.

## 493 KitchenAid

414 N. Peters Rd., Knoxville, TN 37922
Phone: 800 422 1230   Fax: 423 470 6903

**Price Index Number:** 1.88
**Price Index Standard:** standard residential dishwasher

### DISHWASHER
Manufacturer of efficient dishwasher, model #KUDA23B, which uses 6.9 gallons of water for each wash and has an estimated electrical consumption of 695 kwh per year. KitchenAid is a subsidiary of Whirlpool.

## 494 Maytag

1 Dependability Square, Newton, IA 50208
Phone: 515 792 7000 or 515 791 8911
Fax: 515 791 8794

**Price Index Number:** 1.15
**Price Index Standard:** standard residential refrigerator

### REFRIGERATOR
Manufacturer of an efficient, 19-cu.-ft. refrigerator, model #RTD19EODA, with a top-mounted freezer and automatic defrost. Annual operating cost is $52 per year at 8.67¢ per kwh.

## 495 Robert Bosch Corp.

2800 S. 25th Ave., Broadview, IL 60153
Phone: 800 866 2022 or 708 865 5200
Fax: 708 865 5497

**Price Index Number:** 1.5-2.3
**Price Index Standard:** standard residential dishwasher

### DISHWASHER
Manufacturer of efficient dishwashers, models #SMU2000, 3000, 4000, 5000, and 7000 series, which use 5.4 gallons of water for each wash. Tub and inner door made of stainless steel. Come with a 25-year warranty. Made in Germany.

## 496 Robur Corp.

2300 Lynch Rd., Evansville, IN 47711-2908
Phone: 812 424 1800   Fax: 812 422 5117

**Price Index Number:** 2.0, SERVEL
**Price Index Standard:** standard residential refrigerator

### REFRIGERATOR
Manufacturer of SERVEL propane refrigerator which operates on no more than 2 gallons of propane per week.

## 497 Sun Frost

PO Box 95518, Arcata, CA 95521-1101
Phone: 707 822 9095   Fax: 707 822 6213

**Price Index Number:** 3.7-4.2
**Price Index Standard:** standard residential refrigerator

### REFRIGERATOR

Very low-energy-consuming refrigerator. Consumes typically between one-third and one-tenth the energy of a standard refrigerator. Available as 115 VAC, 220 VAC, 12 VDC, or 24 VDC. Also designed for photovoltaics. There are 12 models to choose from, including 19-cu.-ft. refrigerator and/or freezer models. Operates on a quiet short-duty cycle. Designed to function without crisper drawers with separate cooling fins. The compressor is optimized for refrigerator temperature. Modern European styling with recessed handles; may be finished in custom real wood, as well as in a variety of colors. Non-CFC refrigerant (R-134a) refrigerator available.

## 498 Whirlpool

414 N. Peters Rd., Knoxville, TN 37922
Phone: 800 253 1301   Fax: 423 470 6903

**Price Index Number:** 1.5
**Price Index Standard:** standard residential refrigerator

### REFRIGERATOR

Manufacturer of an efficient 22-cu.-ft. refrigerator, which costs approximately $56 per year to operate at 8.30¢ per kwh. It is model #ED22DCXB, a side-by-side refrigerator/freezer developed under the Super Efficient Refrigerator Program. It uses a non-CFC refrigerant, R-134a.

## 499 Coyuchi, Inc.

11101 State Rt. One, Point Reyes Station, CA 94956
Phone: 415 663 9077   Fax: 415 663 8104

**Price Index Number:** 2.0-5.7
**Price Index Standard:** standard cotton wallcovering

### TEXTILES

Manufacturer of environmentally responsible cotton fabrics made from a combination of white cotton and brown and green Foxfiber cotton, all organically grown and free of dyes. Also available are blended yarns of organic cotton and other natural fibers, such as silk, wool, linen, and cashmere. Coyuchi fabrics have been designed in a variety of weave structures for both apparel and interior uses. Color blends include soft sage green, browns in peach and rose tones, olive, and natural white. 15% discount on orders over 50 yards. Weights range from 4 oz. to 15 oz. per sq. yd.

## 500 Design Tex Fabrics

200 Varick St., New York, NY 10014
Phone: 800 797 4949 or 212 886 8100
Fax: 212 886 8149

**Price Index Number:** 5.4
**Price Index Standard:** Naugahyde synthetic leather

### TEXTILES

Environmentally intelligent fabrics for wall coverings, drapery, and upholstery, including the McDonough line of environmental fabrics. WATERSHED is a synthetic leather, water-based polyurethane product, which is biodegradable with the exception of a polyester backing.

## 501  Berkeley Architectural Salvage

722 Folger Ave., Berkeley, CA 94710
Phone: 510 849 2025

**Price Index Number:**  **0.40-0.80**
**Price Index Standard:**  **new products and materials**

**SALVAGED MATERIALS**
Distributor of old recycled building materials and products, such as cabinets, windows, doors, and door hardware. Also available are new recycled products, such as windows purchased on a closeout or as factory seconds.

## 502  Building Futures

2369 University Ave., E. Palo Alto, CA 94303
Phone: 415 473 9838   Fax: 415 473 0913

**Price Index Number:**  **0.20-0.40**
**Price Index Standard:**  **new products and materials**

**SALVAGED MATERIALS**
Nonprofit distributor of salvaged building materials and products, such as lumber, cabinets, windows, doors, and door hardware. Products are listed in the *Recycled Building Materials* newsletter. For information call (415) 856-0634.

## 503  Building Materials Distributors

1708 Cactus Road, San Diego, CA 92173
Phone: 619 661 7181   Fax: 619 661 2571

**Price Index Number:**  **0.25-0.50**
**Price Index Standard:**  **new products and materials**

**SALVAGED MATERIALS**
Demolition contractor and nonprofit distributor of a full range of salvaged building materials and products, such as lumber, cabinets, windows, doors, door hardware, and plumbing fixtures. The company also carries donated and purchased products that are factory closeouts and seconds. Located on ½-acre site in San Diego and another location in Tijuana, Mexico.

## 504  Building Resources

701 Amador St., San Francisco, CA 94124
Phone: 415 285 7814   Fax: 415 285 4689

**Price Index Number:**  **0.25-0.33**
**Price Index Standard:**  **new products and materials**

**SALVAGED MATERIALS**
Nonprofit distributor of a full range of salvaged building materials and products, such as lumber, cabinets, windows, doors, door hardware, and plumbing fixtures. The company also has donated discontinued new products.

## 505  C and M Diversified

330 N. Montgomery, San Jose, CA 95110
Phone: 408 294 5185   Fax: 408 289 9337

**Price Index Number:**  **0.10-0.20**
**Price Index Standard:**  **new products and materials**

**SALVAGED MATERIALS**
Distributor of salvaged building materials and products, such as doors, windows, and plumbing fixtures.

## 506 Caldwell Building Wrecking

195 Bayshore Blvd., San Francisco, CA 94124
Phone: 415 550 6777   Fax: 415 550 0349

**Price Index Number:** **0.20-0.50**
**Price Index Standard:** **new products and materials**

**SALVAGED MATERIALS**
Distributor of salvaged building materials and products, such as lumber and windows. Keeps a large quantity of materials on-site. Also sells a significant quantity of factory seconds and closeouts, such as windows.

## 507 Last Chance Mercantile

PO Box 609, 14201 Del Monte Blvd.,
Marina, CA 93933
Phone: 408 384 5313   Fax: 408 384 3567

**Price Index Number:** **0.50**
**Price Index Standard:** **new products and materials**

**SALVAGED MATERIALS**
Nonprofit distributor of a full range of salvaged building materials and products, such as lumber, cabinets, windows, doors, door hardware, and plumbing fixtures. Also carries donated factory closeouts and seconds.

## 508 Loading Dock

2523 Gwynns Falls Parkway, Baltimore, MD 21216
Phone: 410 728 3625   Fax: 410 728 3633

**Price Index Number:** **0.25-0.33**
**Price Index Standard:** **new products and materials**

**SALVAGED MATERIALS**
Nonprofit distributor of a full range of salvaged building materials and products, such as lumber, cabinets, windows, doors, door hardware, and plumbing fixtures. Sells to nonprofit establishments and low-income households.

## 509 Off The Wall Architectural Antiques

Lincoln between 5th and 6th, Carmel, CA 93921
Phone: 408 624 6165

**Price Index Number:** **0.50-1.0**
**Price Index Standard:** **new products and materials**

**SALVAGED MATERIALS**
Distributor of upscale salvaged architectural antiques, such as stained-glass windows and claw-foot bathtubs.

## 510 Omega Salvage

2407 San Pablo Ave., Berkeley, CA 94702
Phone: 510 843 7368   Fax: 510 548 3552

**Price Index Number:** **0.33**
**Price Index Standard:** **new products and materials**

**SALVAGED MATERIALS**
Distributor of salvaged building materials and products, such as windows, doors, door hardware, stained glass, tile, and plumbing fixtures. Specializes in high-end architectural antiques. Also available are reproductions of antiques.

### 511 Sanger Sales

1355 Felipe Ave., San Jose, CA 95122
Phone: 408 288 5308

**Price Index Number:** **0.30-0.40**
**Price Index Standard:** **new products and materials**

**SALVAGED MATERIALS**
Distributor of salvaged building materials and products, such as lumber, plywood, doors, windows, skylights, cabinets, and bathtubs.

### 512 Urban Ore

1333 6th St., Berkeley, CA 94710
Phone: 510 559 4460 or 510 559 4450 (store)
Fax: 510 528 1540 (store)

**Price Index Number:** **0.30-0.60**
**Price Index Standard:** **new products and materials**

**SALVAGED MATERIALS**
Distributor of salvaged building materials and products, such as cabinets, windows, doors, door hardware, tile, and plumbing fixtures. Maintains a large quantity of items on a 2-acre lot. Specializes in lower-cost building materials and products.

## ART GLASS WORK 12170

### 513 Counter/Productions

705 Bancroft Way, Berkeley, CA 94710
Phone: 510 843 6916   Fax: 510 548 4365

**Price Index Number:** **1.0**
**Price Index Standard:** **granite**

**COUNTERTOP**
Manufacturer of countertops made from 80% recycled glass content. RECYCLED GLASS COUNTERTOPS can consist of thousands of multicolored glass chips surrounded by a solid color binder, up to 2 in. thick. Tested to 5,000 psi compressive strength, the surface is good for cutting, heat, and impact. Finished surface is polished and coated with a penetrating sealer to protect against oils, grease, and acids. Sealer is USDA approved for food contact.

## FURNITURE AND ACCESSORIES 12600

### 514 Creative Office Systems, Inc.

2442 Estand Way, Pleasant Hill, CA 94523
Phone: 510 686 6355   Fax: 510 686 8469

**Price Index Number:** **0.5-0.8**
**Price Index Standard:** **new office furniture**

**REFURBISHED OFFICE FURNITURE**
Manufacturer of office furniture from used products, such as modular panel systems, desks, chairs, bookcases, and work surfaces. Company uses only water-based paints and glue. Space planning and a specification service are included with purchase. Rental, installation, and delivery are also available.

### 515 Herman Miller, Inc.

8500 Byron Rd., Zeeland, MI 49464
Phone: 616 772 3300 or 616 654 8498 (des. & dev.)
Fax: 303 571 4888

**Price Index Number:** **1.0**
**Price Index Standard:** **commercial furniture**

**OFFICE FURNITURE**
Manufacturer of office furniture with recycled content. ERGON and EQUA chairs have 45% recycled content by weight in aluminum and plastic. ACTION OFFICE series work surfaces have 40% recycled content in laminate and particleboard construction. Modular panels have 25% recycled content. All products use only water-based glues and powder-coat paint process to eliminate VOCs and paint waste. Company maintains a research and development program in indoor air quality, recycling, and waste reduction. Old furniture is refurbished by the company's Phoenix Design subsidiary for resale.

## 516 Loewenstein

PO Box 10369, Pompano Beach, FL 33061-6369
Phone: 800 327 2548 or 305 960 1100
Fax: 305 960 0409

**Price Index Number:** 1.0
**Price Index Standard:** commercial furniture

### OFFICE FURNITURE

Manufacturer of office furniture without phenolic resins and HFCs. Only water-based glues are used in products.

## 517 Miller SQA

10875 Chicago Dr., Zeeland, MI 49464
Phone: 800 253 2733 (cust. service)
or 616 772 8000   Fax: 916 632 4317

**Price Index Number:** 0.80
**Price Index Standard:** commercial furniture

### REFURBISHED OFFICE FURNITURE

Herman Miller's ACTION OFFICE series old furniture is refurbished by its Miller SQA subsidiary for resale as a product line called AS NEW. Product line is currently limited to partitions, work surfaces, and lateral files.

## 518 Myers Architectural

555 Basalt Ave., PO Box 1025, Basalt, CO 81621
Phone: 970 927 4761   Fax: 970 927 4610

**Price Index Number:** 1.0
**Price Index Standard:** commercial furniture

### METAL FURNITURE

Manufacturer of custom metal office furniture with 60% recycled content.

## 519 The Natural Bedroom Co.

2585 3rd St., San Francisco, CA 94107
Phone: 800 365 6563 or 415 920 0790
Fax: 415 920 0808

**Price Index Number:** 1.5, organic cotton/ wool blend
**Price Index Standard:** commercial cotton mattresses

### BEDDING AND LINENS

Manufacturer of futon-type mattresses using 80% organically grown cotton and 20% untreated wool. Also distributor of natural comforters, linens, pillows, and other bedding products made from natural fibers without chemical treatments, dyes, or bleaches. Wood bedroom furniture manufactured from sustainable yield sources is also available.

## 520 Royal Pedic

119 N. Fairfax, Los Angeles, CA 90036
Phone: 800 487 6925 or 310 538 9805
Fax: 310 538 9808

**Price Index Number:**    **1.0**
**Price Index Standard:**    **top-quality**
           **polyurethane-filled**
           **mattress**

**BEDDING AND MATTRESSES**
Manufacturer of mattresses using natural materials, such as cotton, wool, and natural latex. No toxic off-gassing or VOCs.

## 521 Vitra

6560 Stonegate Dr., Allentown, PA 18106
Phone: 800 336 9780 or 610 391 9780
Fax: 610 391 9816

**Price Index Number:**    **1.1-1.15**
**Price Index Standard:**    **top-quality office**
           **furniture**

**OFFICE CHAIRS AND TABLES**
Manufacturer of high-quality office work station and waiting-room chairs and tables from recycled polypropylene and aluminum. The company takes an environmental approach when manufacturing and shipping. Products made in Germany. Typical clients include BMW U.S.A.

## 522 Mats, Inc.

37 Shuman Ave., Stoughton, MA 02072
Phone: 617 344 1537

**Price Index Number:**    **1.0**
**Price Index Standard:**    **rubber entry mat**

**MATS**
Entry mats manufactured from recycled tires.

## 523 R. C. Musson

1320 E. Archwood, Akron, OH 44306
Phone: 800 321 2381 or 330 773 7651
Fax: 330 773 3254

**Price Index Number:**    **0.80 (gray)/**
           **1.10 (colors)**
**Price Index Standard:**    **indoor-outdoor**
           **carpet, polypropylene**

**MATS**
DURA RUG and FLUFF CORD are mats manufactured from recycled tires. Available in gray, blue, green, and brown. The mats are spike resistant and come with a 5-year warranty.

## 524 USCOA International Corp.

PO Box 578, St. George, SC 29477
Phone: 800 553 8950 or 803 563 4561
Fax: 803 563 4078

**Price Index Number:**    **0.4, COCO/**
           **0.9, VINA-COIR**
**Price Index Standard:**    **rubber entry mat**

**MATS**
COCO mats are natural-fiber entry mats made from coconut husks. VINA-COIR is a product line with a fused vinyl backing, which is available in larger mat sizes.

### 525 Ashland Rubber Mats Co., Inc.

1221 Elm St., Ashland, OH 44805
Phone: 800 289 1476 or 419 289 7614
Fax: 419 281 7356

**Price Index Number:**  2.0
**Price Index Standard:**  virgin rubber flooring

**WALK-OFF MAT**
ECO-LINK is an open-weave, interlocking floor mat for both residential and commercial applications. Weighs 2.1 lb./sq. ft. Manufactured from 95% recycled rubber, steel, and plastic.

### 526 J Squared Timberworks, Inc.

449 N. 34th St., Seattle, WA 98103
Phone: 800 598 3074 or 206 633 0504
Fax: 206 633 0565

**Price Index Number:**  1.1
**Price Index Standard:**  new wood mantels

**FIREPLACE MANTELS**
Fireplace mantels, surrounds, and custom architectural millwork manufactured from 100% reclaimed timbers and lumber from mill waste, barns, and warehouses.

### 527 Wood Floors, Inc.

PO Box 1522, Orangeburg, SC 29116-1864
Phone: 803 534 8478 or 803 534 3445
Fax: 803 533 0051

**Price Index Number:**  1.0
**Price Index Standard:**  new wood mantels

**FIREPLACE MANTELS**
Fireplace mantels made from 100% reclaimed heart pine lumber salvaged from 18th-century building timbers.

### 528 Aromat Corp.

1935 Lundy Ave., San Jose, CA 95131
Phone: 800 223 6247   Fax: 408 432 9527

**Price Index Number:**  3.0
**Price Index Standard:**  standard tank toilet

**TOILET SEAT**
NAIS Personal Hygiene System is a sophisticated toilet seat attachment that will fit most toilets and offers warm-water cleansing and warm-air drying, reducing or eliminating the consumption of toilet paper. Provides superior comfort and hygiene to tissue cleansing. Requires 120-volt electrical outlet. Has been in use in Japan for 15 years with sales of 1.6 million per year. First introduced in the United States in 1991.

### 529 Clivus Multrum, Inc.

15 Union St., Lawrence, MA 01840
Phone: 800 425 4887 or 508 725 5591
Fax: 508 557 9658

**Price Index Number:**  0.40
**Price Index Standard:**  septic system with
                          flush toilet

**COMPOSTING TOILET**
CLIVUS MULTRUM is a composting toilet available in several models for both commercial and residential use. Manufactured from high-density linear polyethylene, which is recyclable.

## 530 Envirovac, Inc.

1260 Turret Dr., Rockford, IL 61111
Phone: 800 435 6951 or 815 654 8300
Fax: 815 654 8306

**Price Index Number:** **10.0**
**Price Index Standard:** **gravity flush toilet**

### VACUUM TOILET

Vacuum toilet system for water conservation in residential and commercial building applications. Each flush consumes 2 pints of water.

---

## 531 American Standard

PO Box 6820, 1 Centennial Plaza,
Piscataway, NJ 08855
Phone: 800 223 0068, 800 442 1902 (tech)
or 908 980 3000   Fax: 908 980 3335

**Price Index Number:** **0.30-0.40**
**Price Index Standard:** **cast-iron tub**

### BATHTUB

Bathtubs manufactured with a porcelain enameled surface baked to 1800°F and finish bonded to a metal and composite structure requiring one-half the energy consumed to manufacture cast-iron tubs. The weight is one-half that of cast iron. Manufacturer claims the finished product to be nontoxic.

---

## 532 Blanco America, Inc.

1001 Lower Landing Rd., #607,
Blackwood, NJ 08012
Phone: 800 451 5782 or 609 228 3500
Fax: 609 228 7956

**Price Index Number:** **2.7-6.76**
**Price Index Standard:** **garbage disposal,**
**½ hp**

### WASTE MANAGEMENT SYSTEM

Kitchen sink waste disposal systems are included in a line of stainless steel and quartz sinks or as a separate item. Available with a disposal chute for composting. Many styles and colors available.

---

## 533 Kohler Co.

444 Highland Dr., Kohler, WI 53044
Phone: 800 456 4537 or 414 457 4441
Fax: 414 457 9064

**Price Index Number:** **5.0**
**Price Index Standard:** **standard residential**
**sink**

### WASTE MANAGEMENT SYSTEM

ECO-CYCLE is a kitchen sink waste disposal system consisting of a cast-iron double sink measuring 43 in. by 22 in. and incorporating a built-in disposal chute for composting vegetation waste.

## 534 Advanced Tech Industries

7441 N.W. 8th St., Suite J, Miami, FL 33126
Phone: 305 265 7751   Fax: 305 265 7184

**Price Index Number:** **1.0**
**Price Index Standard:** **residential gas tank water heater**

### TANKLESS WATER HEATER

Manufacturer of an instantaneous water heater that supplies hot water upon demand with a line sensor and chip that optimizes power consumption based on usage. Has a manual temperature control. This system avoids tank and line losses, and requires only a single cold-water supply line. Available in 2 heater sizes for electric 220 volts. The SUPREME S220 provides 4 gallons per minute for whole-house applications, and the SUPREME I220 provides 2.5 gallons per minute for limited-fixture applications. Heater measures 12 in. by 11 in. by 3 in. Claims to achieve 50% savings on tank water heater utility bills and 30% savings over nonelectric instantaneous heaters. Comes with a limited lifetime warranty.

## 535 American Water Heater

PO Box 4056, 500 Princeton Rd.,
Johnson City, TN 37601
Phone: 800 937 1037 or 615 283 8000
Fax: 800 581 7224

**Price Index Number:** **0.70**
**Price Index Standard:** **residential gas tank water heater**

### MINI TANK WATER HEATER

Manufacturer of a miniature electric tank water heater. The TINY TITAN reduces tank and pipe losses by acting as a stand-alone water heater for hand washing. Available in 2- and 2½-gallon sizes with a 7-gallon first hour rating.

## 536 Bradford White

323 Norristown Rd., Ambler, PA 19002
Phone: 800 538 2020   Fax: 215 641 1670

**Price Index Number:** **0.50**
**Price Index Standard:** **residential gas tank water heater**

### MINI TANK WATER HEATER

Manufacturer of miniature electric tank water heater. The POWERFUL COMPACT reduces tank and pipe losses by acting as a stand-alone water heater for hand washing. Available in a 2-gallon size with a 9-gallon first-hour rating.

## 537 Chronomite Labs

1420 W. 240th St., Harbor City, CA 90710
Phone: 800 447 4962 or 310 524 2300
Fax: 310 530 1381

**Price Index Number:** **1.3-2.5**
**Price Index Standard:** **residential gas tank water heater**

### TANKLESS WATER HEATER

Manufacturer of INSTANT-FLOW series instantaneous water heaters that supply hot water upon demand. This system avoids tank and line losses, and requires only a single cold-water supply line. Available in 7 electric heater sizes for 110 volts or 220 volts.

## 538 Controlled Energy Corp.

Fiddler's Green, Waitsfield, VT 05673
Phone: 800 642 3199, 800 642 3111
or 802 496 4436   Fax: 802 496 6924

**Price Index Number:**   **2.5-3.0, AQUASTAR**
**Price Index Standard:**   **residential gas tank**
   **water heater**

### TANKLESS WATER HEATER

Manufacturer of instantaneous water heaters that supply hot water upon demand using a minimum-flow activation sensor. This system avoids tank and line losses, and requires only a single cold-water supply line. Available in 3 heater sizes for electric 120 volts, 240 volts, or gas. The POWERSTREAM series is electric and is designed for commercial hand-washing applications with an adjustable heat-setting selector. It's about the size of a cigar box. The AQUASTAR 80, 125, or 170 are gas units applicable to residential use with an adjustable thermostatic control. Heaters measure 27½ in. by 17 in. by 9½ in.

## 539 Controlled Energy Corp.

Fiddler's Green, Waitsfield, VT 05673
Phone: 800 642 3199 or 802 496 4436
Fax: 802 496 6924

**Price Index Number:**   **0.60-0.70**
**Price Index Standard:**   **residential gas tank**
   **water heater**

### MINI TANK WATER HEATER

Manufacturer of miniature tank water heaters. This heater reduces tank and pipe losses by acting as a stand-alone water heater for hand washing. Available in two sizes for electric 120 volts with an adjustable thermostatic control. The ARISTON P10S has a 2.5-gallon tank, and the ARISTON P15S has a 4-gallon tank.

## 540 EEMAX, Inc.

472 Pepper St., Monroe, CT 06468
Phone: 800 543 6163 or 203 261 0684
Fax: 203 261 4790

**Price Index Number:**   **0.75-3.5**
**Price Index Standard:**   **residential gas tank**
   **water heater**

### TANKLESS WATER HEATER

Manufacturer of instantaneous water heaters that supply hot water upon demand. Electric heaters are for 240 volts.

## 541 In-Sink-Erator

4700 21st St., Racine, WI 53406
Phone: 800 558 5712 or 414 554 5432
Fax: 414 554 8917

**Price Index Number:**   **0.56-0.61**
**Price Index Standard:**   **residential gas tank**
   **water heater**

### MINI TANK WATER HEATER

Manufacturer of miniature electric tank water heaters. The models W-152 and W-154 reduce tank and pipe losses by acting as stand-alone water heaters for hand washing. Available in 2- and 4½-gallon sizes with an 8.4-gallon first-hour rating.

## 542 Keltech, Inc.

PO Box 405, Richland, MI 49083
Phone: 800 999 4320 or 616 629 4814
Fax: 616 629 4853

**Price Index Number:**   **3.0-3.5**
**Price Index Standard:**   **residential gas tank**
   **water heater**

### TANKLESS WATER HEATER

Manufacturer of instantaneous water heaters that supply hot water upon demand. Provided with a remote temperature dial that can be mounted up to 25 ft. from the unit and extended to 75 ft. Available in 6 models, ACUTEMP 100, 120, 150, 153, 180, or 183 electric for 240 volts, single or three phase.

## 543 Kilo Alpha Co.

PO Box 768, Chatham, MA 02633
Phone: 508 945 4747   Fax: 508 945 4790

**Price Index Number:**   **1.0**
**Price Index Standard:**   **instantaneous electric water heater**

### TANKLESS WATER HEATER
Manufacturer of electric pipe water heater that maintains a constant temperature of hot water in pipes. Plugs into a standard 110-volt outlet. This product does the same as a recirculating water heating system, but more efficiently by safely delivering ⅓ volt and 1,000 amperes directly to a copper pipe to produce 100 watts of water heating.

## 544 Rheem/Ruud

PO Box 244020, Montgomery, AL 53406
Phone: 800 432 8373, 800 621 5622,
or 334 260 1500   Fax: 334 260 1341

**Price Index Number:**   **1.0, 2½-gallon size**
**Price Index Standard:**   **residential gas tank water heater**

### MINI TANK WATER HEATER
Manufacturer of miniature electric tank water heater. The POINT-OF-USE-MISER reduces tank and pipe losses by acting as a stand-alone water heater for hand washing. Available in a 2½-gallon size with an 8.4-gallon first-hour rating. Also available in 6-, 10-, 15-, 20-, and 30-gallon capacities from 16 in. to 30 in. high. Has a 5-year limited tank warranty and a 1-year limited parts warranty.

## 545 Rheem/Ruud

PO Box 244020, Montgomery, AL 53406
Phone: 800 432 8373, 800 621 5622,
or 334 260 1500   Fax: 334 260 1341

**Price Index Number:**   **4.0, 50-gallon size**
**Price Index Standard:**   **residential gas tank water heater**

### WATER HEATER
MARATHON/PERFORMER is a high-efficiency tank water heater with an Energy Factor of 70. Seamless polybutylene inner tank is reinforced with fiberglass and is completely surrounded by urethane insulation, which has an R-value of R-16.7. Available in 40- and 50-gallon sizes.

## 546 SPS Marine

25820 Orchard Lake Rd, Suite 3,
Farmington Hills, MI 483336
Phone: 810 476 0531   Fax: 810 476 1918

**Price Index Number:**   **3.5**
**Price Index Standard:**   **residential gas tank water heater**

### TANKLESS WATER HEATER
Manufacturer of an instantaneous water heater that supplies hot water upon demand. The model 300 is a small-capacity, gas-type heater made of stainless steel for marine locations. Typical applications are houseboats and yachts.

## 547 State Industries

500 By Pass Rd., Ashland, TN 37015
Phone: 800 365 0024   Fax: 615 792 2186

**Price Index Number:**   **1.4**
**Price Index Standard:**   **residential gas tank water heater**

### MINI TANK WATER HEATER
Manufacturer of miniature electric tank water heater. The PV-6-10UFK reduces tank and pipe losses by acting as a stand-alone water heater for hand washing. Available in a 6-gallon size with a 7-gallon first-hour rating.

## 548 Therma-Stor Products

PO Box 8050, Madison, WI 53708
Phone: 800 533 7533 or 608 222 5301
Fax: 608 222 1447

**Price Index Number:**   **8.0**
**Price Index Standard:**   **residential tank water heater**

### HEAT PUMP WATER HEATER
THERMASTER HP-ST air-to-water heat pump water heaters. Typical COPs are 2.1 to 2.5.

---

## 549 Vaillant Corp.

2607 River Rd., Cinnaminson, NJ 08077
Phone: 609 786 2000   Fax: 609 786 8465

**Price Index Number:**   **2.6**
**Price Index Standard:**   **residential gas tank water heater**

### TANKLESS WATER HEATER
Manufacturer of an instantaneous water heater that supplies hot water upon demand. The MAG 325 is a gas-type heater with a piezo igniter and an automatically modulating gas control for the amount of hot water being used.

---

## 550 American Standard

PO Box 6820, 1 Centennial Plaza,
Piscataway, NJ 08855
Phone: 800 223 0068, 800 442 1902 (tech)
or 908 980 3000   Fax: 908 980 3335

**Price Index Number:**   **2.2**
**Price Index Standard:**   **standard residential furnace**

### FURNACE
FREEDOM 90 is a gas-fired furnace with a single-stage burner and a single- or variable-speed blower in 11 models from 38,000 to 111,000 BTUH and 91% to 92% efficiency. Dimensions are 28 in. by 17½ in. to 24½ in. by 40 in. high. Available in up-flow, down-flow, and horizontal-flow positions. Comes with a limited lifetime warranty.

---

## 551 Bryant, Day and Night, Payne

PO Box 70, Indianapolis, IN 46206
Phone: 800 468 7253 or 800 227 7437
Fax: 315 432 6620

**Price Index Number:**   **2.0**
**Price Index Standard:**   **standard residential furnace**

### FURNACE
PLUS 90i is a gas-fired furnace with a two-stage burner, sealed combustion air, and variable-speed blower in 4 models from 40,000 to 100,000 BTUH and up to 96.6% efficiency. Dimensions are 28 in. by 17½ in. to 24½ in. by 40 in. high. Available in up-flow, down-flow, and horizontal-flow positions. Comes with a limited lifetime warranty.

## 552 Burnham Corp.

PO Box 3079, Lancaster, PA 17604
Phone: 717 397 4701   Fax: 717 293 5832

**Price Index Number:**   **2.0, boiler only**
**Price Index Standard:**   **standard residential furnace**

### BOILER

XG2000A series induced-draft, gas-fired boilers come in 4 models from 62,000 to 164,000 BTUH and 83.3% to 84.8% efficiency. Dimensions are 21 in. by 16 in. by 33 in. high. SPIRIT HOME COMFORT SYSTEM series are direct-vent boilers with a combined water heating option in 4 models. Capacity from 52,000 to 137,000 BTUH and 83% to 84% efficiency. Dimensions of boilers vary from 20¾ in. to 30½ in. by 24 in. wide by 36 in. high. V1 series natural-draft, oil-fired boilers come in 8 models from 73,000 to 248,000 BTUH and 83.2% to 85.4% efficiency. Dimensions are 21 in. by 15½ in. by 29 in. high. All boilers have cast-iron heat exchangers. Come with a lifetime limited warranty and 5- or 10-year blanket coverage.

## 553 Carrier Corp.

PO Box 4808, Syracuse, NY 13221
Phone: 800 227 7437 or 315 432 6000
Fax: 315 432 6620

**Price Index Number:**   **2.0**
**Price Index Standard:**   **standard residential furnace**

### FURNACE

58MVP series gas-fired furnaces with two-stage burner, sealed combustion air, and variable-speed blower come in 4 models from 40,000 to 100,000 BTUH and up to 96.6% efficiency. Dimensions are 28 in. by 17½ in. to 24½ in. by 40 in. high. Available in up-flow, down-flow, and horizontal-flow positions. The furnaces come with a limited lifetime warranty.

## 554 Glowcore

PO Box 360591, Cleveland, OH 44136
Phone: 800 676 4546   Fax: 216 225 3134

**Price Index Number:**   **4.5**
**Price Index Standard:**   **standard residential furnace**

### BOILER

GB series induced-draft, condensing-type gas-fired boilers come in 5 models from 40,000 to 160,000 BTUH and 92.5% to 94.1% efficiency. Dimensions are 18 in. to 21 in. by 11 in. by 36 in. to 42 in. high.

## 555 Hydrotherm Corp.

260 N. Elm St., Westfield, MA 01085
Phone: 413 568 9571   Fax: 413 568 9613

**Price Index Number:**   **3.7**
**Price Index Standard:**   **standard residential furnace**

### BOILER

MULTIPULSE sealed-combustion, condensing gas boiler is 90% efficient. Boiler is designed to be combined in modules with heat exchangers for optimum efficiency for space heating and domestic hot water.

## 556 Lennox Corp.

PO Box 799900, Dallas, TX 75379
Phone: 800 453 6669 or 972 497 5000
Fax: 972 497 5490

**Price Index Number:**   **2.5-3.0, COMPLETE HEAT**
**Price Index Standard:**   **residential furnace + water heater**

### BOILER

COMPLETE HEAT system is a combination space-heating and water-heating system consisting of a boiler, hot-water tank, and air handler. Efficiency is 90%, and the 30-gallon hot-water tank has very fast recovery, producing at least three times more hot water than conventional water heaters. It comes with a single-stage burner, sealed combustion air, and single-speed blower in 7 models from 40,000 to 120,000 BTUH. Dimensions are 46 in. by 16¼ in. to 21¼ in. by 28½ in. Comes with a limited 15-year warranty on the heat exchanger and 5 years on parts.

## 557 Lennox Corp.

PO Box 799900, Dallas, TX 75379
Phone: 800 453 6669 or 972 497 5000
Fax: 972 497 5490

**Price Index Number:** 2.5, G-21V/2.1, G-21/
1.5, G-26
**Price Index Standard:** standard residential
furnace

### FURNACE

G-21 PULSE series gas-fired furnaces with single-stage burner, sealed combustion air, and single-speed blower come in 4 models from 55,000 to 95,000 BTUH and 91% to 96% efficiencies. Dimensions are 26 in. by 12¼ in. to 26½ in. by 49 in. to 54¼ in. Available in up-flow, down-flow, and horizontal-flow positions. Comes with a limited lifetime warranty. Also available with a high-efficiency, variable-speed blower in the G-21V. G-26 induced-draft series is the least expensive of the 90+% high-efficiency Lennox furnaces.

## 558 Lochinvar Corp.

2005 Elm Hill Pike, Nashville, TN 37210
Phone: 615 889 8900   Fax. 615 885 4403

**Price Index Number:** 6.3
**Price Index Standard:** standard residential
furnace

### BOILER

PBN-0250 condensing, sealed-combustion gas-fired boiler has a 250,000 BTUH capacity and 88% efficiency. Dimensions are 23 in. by 36 in. by 34 in. high.

## 559 Robur Corp.

2300 Lynch Rd., Evansville, IN 47711
Phone: 812 424 1800 x222   Fax: 812 422 5117

**Price Index Number:** 3.0
**Price Index Standard:** residential heat pump

### CHILLER/HEATER

Manufacturer of gas-fired residential and light commercial absorption chiller/heater. Available as a combination chiller and boiler system for 3- to 5-ton loads. Refrigerant fluid is in a sealed system consisting of 1 part ammonia and 2 parts water with sodium chromate, a toxic chemical, as an inhibitor. Nevertheless no CFCs are used. This is a hydronic system with a chiller COP of 0.62 and a boiler AFUE of 80% to 82%.

## 560 Sanyo-Fisher USA Corp.

21350 Lassen St., Chatsworth, CA 91311
Phone: 818 998 7322   Fax: 818 701 4170

**Price Index Number:** 3.2
**Price Index Standard:** standard residential
furnace with A/C

### MULTIZONE DUCTLESS HEAT PUMP

Manufacturer of multizone, ductless heat-pump systems using wall- or ceiling-mounted terminal units. Output capacities vary from 9,000 to 43,000 BTUH with SEER/HSPF from 9.5 to 11.4/6.6 to 7.3, respectively. Controls can be wired or remote hand-held wireless. Equipped with a 24-hour programmable timer and night setback mode. Auto-louver mechanism moves louvers up and down for optimum air distribution. Each system is limited to either heating or cooling mode at any given time.

## 561 Slant/Fin Corp.

100 Forest Dr., Greenvale, NY 11548
Phone: 800 873 4346 or 516 484 2600
Fax: 516 484 6958

**Price Index Number:** **1.9, VICTORY/ 2.0, LIBERTY**
**Price Index Standard:** **standard residential furnace**

### BOILER
VICTORY induced-draft, gas-fired boilers come in 6 models from 33,000 to 180,000 BTUH and 82.5% to 84.8% efficiency. Dimensions are 19¾ in. by 14⅝ in. by 28⅜ in. high. LIBERTY natural-draft, oil-fired boilers come in 8 models from 91,000 to 329,000 BTUH and 83.3% to 86% efficiency. Dimensions are 24¼ in. to 41⅛ in. by 25 in. by 31⅞ in. high.

## 562 Tadiran Electrical Appliances

575 Lexington Ave., 21st Floor,
New York, NY 10022
Phone: 800 823 7754   Fax: 212 751 1923

**Price Index Number:** **1.5**
**Price Index Standard:** **standard residential furnace with A/C**

### MULTIZONE DUCTLESS HEAT PUMP
Manufacturer of multizone ductless heat-pump systems using wall- or ceiling-mounted terminal units. Output capacities vary from 5,200 to 36,000 BTUH with SEER/HSPF from 10 to 12/6.8 to 7.1, respectively. Controls can be wired or remote hand-held wireless. Two zones maximum for each outside condensing unit. This system has a unique feature of allowing one zone to heat while the other zone may be cooling, with the same condenser.

## 563 Teledyne Laars

20 Industrial Way, Rochester, NH 03867
Phone: 800 362 5678 or 603 335 6300
Fax: 603 335 3355

**Price Index Number:** **1.6**
**Price Index Standard:** **standard residential furnace**

### BOILER
MINI THERM II series natural-draft, gas-fired boilers come in 4 models from 50,000 to 125,000 BTUH and 83% to 84.4% efficiency. Dimensions are 24 in. by 28 in. by 59¾ in. high.

---

**HEAT TRANSFER** | 15750

## 564 Earthstar Energy Systems

PO Box 626, Waldoboro, MA 04572
Phone: 800 323 6749 or 207 832 6861
Fax: 207 832 7314

**Price Index Number:** **2.5**
**Price Index Standard:** **standard residential water heater**

### WATER HEAT EXCHANGER
GREYWATER HEAT RECLAIMER is a cross-linked polyethylene heat exchanger in a 50-gallon seamless polyethylene tank designed to recover heat from gray water to preheat water entering a domestic hot-water tank. Manufacturer claims an average annual savings on water heating bills to be 40%.

## 565 Honeywell, Inc.

1985 Douglas Dr. N., Golden Valley, MN 55422
Phone: 800 468 1502 or 612 951 1000
Fax: 612 951 2086

**Price Index Number:** **2.6-3.6, furnace + AIRXCHANGE**
**Price Index Standard:** **standard residential furnace**

### HEAT-EXCHANGER VENTILATOR
AIRXCHANGE is a counter-flow, heat-exchanger ventilator for recovering space heat from vented air in residential and light commercial applications. Also available as a combination heat-exchanger ventilator and fan coil package for use with domestic water heater. Water heater must be a minimum of 50 gallons with a capacity of 100,000 BTUH.

## 566 Nutech

511 McCormick Blvd., London, ON,
Canada N5W 4C8
Phone: 519 457 1904   Fax: 519 457 1676

**Price Index Number:**   **1.6**
**Price Index Standard:**   **electrostatic air filter**

### HEAT-EXCHANGER VENTILATOR

LIFE BREATH heat-recovery ventilator (HRV) is a self-contained unit measuring 15 in. by 19 in. by 31 in. with 2 input and 2 output duct connections, which may be incorporated into a forced-air system in several ways. This product includes programmable humidity control, timers, and air-quality sensor.

## 567 Therma-Stor Products

PO Box 8050, Madison, WI 53708
Phone: 800 533 7533 or 608 222 5301
Fax: 608 222 1447

**Price Index Number:**   **2.2, THERMA VENT/
3.0, ENVIRO VENT**
**Price Index Standard:**   **residential furnace
+ water heater**

### HEAT-EXCHANGER VENTILATOR

THERMA VENT and ENVIRO VENT heat-recovery ventilators come with a heat pump to extract heat from bathroom air for water or space heating. SAHARA whole-house air system combines heat recovery, humidity control, and a 7-day timer for air-quality control and energy savings.

## 568 Vent-Aire Systems

4850 Northpark Dr., Colorado Springs, CO 80918
Phone: 800 937 9080 or 719 599 9080
Fax: 719 599 9085

**Price Index Number:**   **0.9**
**Price Index Standard:**   **heat-exchanger
ventilator and fan coil**

### HEAT-EXCHANGER VENTILATOR

Manufacturer of counter-flow heat-exchanger ventilator for recovering space heat from vented air in residential and light commercial applications. Also available as a combination heat-exchanger ventilator and fan coil package for use with domestic water heater. Water heater must be a minimum of 50 gallons with a capacity of 100,000 BTUH.

## 569 EHT Siegmund, Inc.

14771 Myford Rd., Suite C, Tustin, CA 92680
Phone: 800 704 6016 or 714 731 4143
Fax: 714 731 3018

**Price Index Number:**   **2.0**
**Price Index Standard:**   **hydronic floor
heating system**

### RADIANT FLOOR HEATING AND COOLING

Hydronic radiant floor heating and cooling system with cross-linked polyethylene tubing.

## 570 Stadler Corp.

3 Alfred Circle, Bedford, MA 01730
Phone: 800 370 3122 or 617 275 3122
Fax: 617 275 5398

**Price Index Number:**   **1.0, material**
**Price Index Standard:**   **copper tubing**

### RADIANT FLOOR HYDRONIC SYSTEM

Radiant floor hot-water systems with individual room zones. Available floor tubing is PEXTRON, a cross-linked polyethylene tubing for hot-water floor systems with an oxygen diffusion barrier. Available in ½-in., ¾-in., and 1-in. diameters.

### 571  ThermalEase

20714 State Rt. 305, Suite 3C, Poulsbo, WA 98370
Phone: 360 779 1960   Fax: 360 779 1343

**Price Index Number:   0.60-1.08, material**
**Price Index Standard:   copper tubing**

**RADIANT FLOOR HYDRONIC SYSTEM**
Radiant floor hot-water systems with unlimited zones. Available floor tubings are OXYGUARD BARRIER PEX, a cross-linked polyethylene tubing for hot-water floor systems with an oxygen diffusion barrier, and OXYGUARD BARRIER PB, which is polybutylene with an oxygen barrier. PB and PEX are available in ⅜-in., ½-in., ¾-in., and 1-in. diameters, and also ⅝-in. in PEX. PEX is manufactured to ASTM F-876, F-877, and CSA B-137.5 in ISO 9002 certified factory.

### 572  Wirsbo Co.

5925 148th St. W., Apple Valley, MN 55124
Phone: 800 321 4739 or 612 891 2000
Fax: 612 891 1246

**Price Index Number:   0.90, material/**
**0.40, material + labor**
**Price Index Standard:   copper tubing**

**RADIANT FLOOR HYDRONIC SYSTEM**
Radiant floor hot-water systems with unlimited zones. Available floor tubings are HEPEX, a cross-linked polyethylene tubing for hot water floor systems with an oxygen diffusion barrier, and PEX, which is the same as HEPEX without the oxygen barrier. Available in ⅜-in., ½-in., ⅝-in., ¾-in., and 1-in. diameters.

## AIR-CLEANING DEVICES

**15885**

### 573  Nutech

511 McCormick Blvd., London, ON,
Canada N5W 4C8
Phone: 519 457 1904   Fax: 519 457 1676

**Price Index Number:   1.6**
**Price Index Standard:   electrostatic air filter**

**AIR FILTER**
LIFEBREATH turbulent-flow precipitation (TFP) air filter is a new concept in high-efficiency, low-maintenance air filtration. The manufacturer claims that the product will equal the performance (94% efficiency) of an electrostatic filter for 4 years without maintenance. AIR SENTRY air-quality sensor uses a solid-state gas detector to sense the presence of air pollution. If pollution is detected, a display flashes and increases the ventilation rate for the system.

## LIGHTING

**16500**

### 574  Aero-Tech Light Bulb Co.

534 Pratt Ave. N., Schaumburg, IL 60193
Phone: 847 352 4900   Fax: 847 352 4999

**Price Index Number:   20-30**
**Price Index Standard:   incandescent lamp**

**COMPACT FLUORESCENT LAMPS**
Manufacturer of modular-type compact fluorescent bulbs using magnetic ballasts. Available as double tube and floodlights in 9 watts and 13 watts. These bulbs use 70% less electricity than incandescent bulbs and last up to 10,000 hours.

## 575 Beacon Light Products, Inc.

723 W. Taylor Ave., Meridian, ID 83642
Phone: 208 888 5905   Fax: 208 888 7433

**Price Index Number:** 9.0, bulb +
computerized module
**Price Index Standard:** incandescent lamp

### LIGHTING CONTROLS

BULB BOSS is an integrated circuit light control module, about the size of a dime, which is placed in a light socket to perform preprogrammed functions that control the performance of an incandescent or halogen bulb. Available in 6 modules: automatic preset 10-minute and 30-minute off switches with an override; 4-way dimmer; emergency flasher; night light that gradually dims over a 20-minute period; and a 6-hour on and 18-hour off cycle timer. All modules are programmed to protect the bulb from the initial surge of power when it is turned on. A soft start feature extends the life of a bulb 2 to 4 times, according to the manufacturer. BULB BOSS is expected to last up to 20 years. It comes with a 1-year limited warranty.

## 576 Energy Solutions International

11003 I St., Omaha, NE 68137
Phone: 402 592 2363   Fax: 402 592 2570

**Price Index Number:** bid on job-by-
job basis

### EFFICIENT LIGHTING

3M SILVERLUX and aluminum reflectors are used in customized fluorescent fixtures for energy-efficient lighting. The company provides manufacturing, installation, and servicing of fixtures. The manufacturer is able to convert a standard 2-ft. by 4-ft. fluorescent fixture consuming 192 watts into a 62-watt fixture while maintaining the same lighting level.

## 577 General Electric Co.

1975 Noble Rd., Cleveland, OH 44112
Phone: 800 435 4448   Fax: 800 327 0663

**Price Index Number:** 20-30
**Price Index Standard:** incandescent lamp

### COMPACT FLUORESCENT LAMPS

Manufacturer of integral- and modular-type compact fluorescent bulbs using electronic or magnetic ballasts. Available as single, double, triple, or quad tube, globe, tubular, circular, U-shaped, and reflector lamps in 5, 7, 13, 15, 18, 20, 22, 23, 24, 25, 26, 27, 28, and 39 watts. These bulbs use 70% less electricity than incandescent bulbs and last up to 10,000 hours.

## 578 Maxlite SK America, Inc.

60 E. Commerce Way, Totowa, NJ 07512
Phone: 800 793 1212 or 201 256 3330
Fax: 201 256 9444

**Price Index Number:** 20-30
**Price Index Standard:** incandescent lamp

### COMPACT FLUORESCENT LAMPS

Manufacturer of modular- and integral-type compact fluorescent bulbs using electronic or magnetic ballasts. Available as reflector, globe, circular, and tubular lamps in 15, 17, 18, 20, 23, and 30 watts. Colored bulbs are available in red, green, and blue.

## 579 Osram Sylvania

100 Endicott St., Danvers, MA 01923
Phone: 508 777 1900   Fax: 508 750 2152

**Price Index Number:** 20-30
**Price Index Standard:** incandescent lamp

### COMPACT FLUORESCENT LAMPS

Manufacturer of modular- and integral-type compact fluorescent bulbs using electronic or magnetic ballasts. Available as double or triple tube, globe, and reflector lamps in 9 watts and 13 watts. These bulbs use 70% less electricity than incandescent bulbs and last up to 10,000 hours.

## 580  Panasonic Lighting Co.

One Panasonic Way, Secaucus, NJ 07094
Phone: 714 373 7164 (West) or
201 348 5381 (East)
Fax: 800 842 1031 (West) or 800 553 0384 (East)

**Price Index Number:    20-30**
**Price Index Standard:   incandescent lamp**

**COMPACT FLUORESCENT LAMPS**
Manufacturer of integral-type compact fluorescent bulbs
using electronic ballasts. Available as double tube, globe, and
tubular lamps in 15, 16, 20, and 25 watts. These bulbs use
70% less electricity than incandescent bulbs and last up to
10,000 hours.

## 581  Philips Lighting Co.

PO Box 6800, Somerset, NJ 08875
Phone: 908 563 3000    Fax: 908 563 3641

**Price Index Number:    20-30**
**Price Index Standard:   incandescent lamp**

**COMPACT FLUORESCENT LAMPS**
Manufacturer of integral-type compact fluorescent bulbs
using electronic ballasts. Available as double or triple tube,
reflector, and tubular lamps in 9, 11, 15, 17, 18, 20, and 23
watts. These bulbs use 70% less electricity than incandescent
bulbs and last up to 10,000 hours.

## 582  Supreme Lighting Co.

1605 John St., Fort Lee, NJ 07024
Phone: 800 221 1573    Fax: 201 947 9329

**Price Index Number:    20-30**
**Price Index Standard:   incandescent lamp**

**COMPACT FLUORESCENT LAMPS**
Manufacturer of integral-type compact fluorescent bulbs
using magnetic ballasts. Available as double tube and
floodlight lamps in 9, 13, and 16 watts. Also has conversion
kits with wall-outlet and socket adapter using magnetic
ballasts for double tubes in 7, 9, and 13 watts. These bulbs
use 70% less electricity than incandescent bulbs and last up
to 10,000 hours.

## 583  US-PAR Enterprises

13404 S. Monte Vista Ave., Chino, CA 91710
Phone: 909 591 7506    Fax: 909 590 3220

**Price Index Number:    20-30**
**Price Index Standard:   incandescent lamp**

**COMPACT FLUORESCENT LAMPS**
Manufacturer of integral-type compact fluorescent bulbs
using magnetic ballasts. Available as double tube, floodlight,
globe, circular, and U-shaped lamps in 5, 7, 9, 12, 13, 15, 22,
and 30 watts. These bulbs use 70% less electricity than
incandescent bulbs and last up to 10,000 hours.

# MANUFACTURER INDEX

| Manufacturer Name | Product Category | Record Number |
|---|---|---|

## A

| | | |
|---|---|---|
| Advanced Environmental Recycling Technologies | Wood/Resin Composite | 148 |
| Advanced Foam Plastics, Inc. | Concrete Form System | 47 |
| | Expanded Polystyrene | 203 |
| Advanced Framing Systems, Inc. | Steel Framing | 76 |
| Advanced Tech Industries | Tankless Water Heater | 534 |
| Aeolian Enterprises | Plastic Lumber | 169 |
| Aero-Tech Light Bulb Co. | Compact Fluorescent Lamps | 574 |
| AFM Corp. | Manufactured Wall Panels | 281 |
| AFM Enterprises, Inc. | Adhesives | 359 |
| | Caulk and Putty | 314 |
| | Duct Mastic | 317 |
| | Paints and Stains | 440 |
| | Waterproofing | 196 |
| Aged Woods, Inc. | Recycled Lumber | 79 |
| Agriboard Industries | Manufactured Wall Panels | 282 |
| Alaska Window Co. | Vinyl Windows | 331 |
| Albany Woodworks | Recycled Flooring | 374 |
| Alenco | Vinyl Windows | 332 |
| Alket Industries | Plastic Lumber | 170 |
| All Fiberglass Products Corp. | Fence Screening | 29 |
| All-Weather Insulation Co. | Loose-Fill Insulation | 204 |
| Allegro Rug Weaving | Area Rugs | 414 |
| Allied Demolition, Inc. | Recycled Lumber | 80 |
| Alpha Granite and Marble | Tile and Sheet Flooring | 389 |
| Alside | Vinyl Windows | 333 |
| Amana Refrigeration | Refrigerator | 489 |
| Amazing Recycled Products | Plastic Lumber | 171 |
| | Site Traffic Control | 5 |
| American Cemwood Products | Fiber-Cement Shingles | 263 |
| American ConForm Industries, Inc. | Polystyrene Concrete Form System | 48 |
| American Insulation, Inc. | Loose-Fill Insulation | 205 |
| American Polysteel Forms | Polystyrene Concrete Form System | 49 |
| American Rockwool, Inc. | Mineral Wool Insulation | 206 |
| American Sprayed Fibers, Inc. | Acoustical Insulation | 368 |
| | Foam Insulation | 207 |
| | Spray Fireproofing | 262 |
| American Standard | Bathtub | 531 |
| | Furnace | 550 |
| American Water Heater | Mini Tank Water Heater | 535 |
| AMREC | Asphaltic Concrete Paving | 6 |
| Amtico International | Vinyl Flooring | 390 |
| Andersen Windows | Wood Windows | 334 |
| Angeles Metal Systems | Steel Studs | 73 |
| Applegate Manufacturing | Loose-Fill Insulation | 208, 209 |
| AquaPore Moisture Systems | Drip Irrigation Hose | 28 |
| Aquazone Products Co. | Ozone Purification System | 480 |
| Architectural Forest Enterprises | Fiberboard | 149 |
| Arctic Manufacturing | Loose-Fill Insulation | 210 |
| Ark-Seal, Inc. International | Sprayed Insulation | 211 |
| Armstrong World Industries | Acoustical Ceilings | 369 |
| Aromat Corp. | Toilet Seat | 528 |
| Ashland Rubber Mats Co., Inc. | Walk-off Mat | 525 |
| Asko, Inc. | Dishwasher | 490 |
| ATAS International, Inc. | Metal Shingles | 264, 265 |
| Atlas Roofing Corp. | Roofing Shingles | 266 |

| Manufacturer Name | Product Category | Record Number |
|---|---|---|
| Auro-Sinan Co. | Caulk and Putty | 315 |
| | Paints | 441 |
| | Plaster/Joint Compound | 354 |

## B

| Manufacturer Name | Product Category | Record Number |
|---|---|---|
| Barrier Technology USA | Fire-Rated Sheathing | 160 |
| Bass and Hays Foundry, Inc. | Gutters, Leaders, and Drains | 312 |
| BathCrest, Inc. | Porcelain Surface | 438 |
| Beacon Light Products, Inc. | Lighting Controls | 575 |
| Bear Creek Lumber Co. | Sustainable Harvest | 81 |
| Bedard Cascades, Inc. | Roofing Felt | 306 |
| Bedford Industries | Plastic Lumber | 172 |
| Bedrock Industries | Ceramic Tile | 361 |
| | Rubble Stone Flooring Tile | 360 |
| Bellcomb Technologies | Manufactured Wall Panels | 283 |
| Benjamin Moore and Co. | Paints | 442 |
| Berkeley Architectural Salvage | Salvaged Materials | 501 |
| Bestmann Green Systems, Inc. | Soil Stabilization | 1 |
| Big Creek Lumber Co. | Sustainable Harvest | 82 |
| BioFab | Particleboard | 150 |
| Biofire, Inc. | Masonry Fireplaces | 467 |
| Blanco America, Inc. | Waste Management System | 532 |
| Bloomsburg Carpet Industries, Inc. | Carpet | 415 |
| Blue Ridge Carpet Mills | Carpet | 416 |
| BMCA Insulation Products, Inc. | Board Insulation | 212 |
| Boise Cascade Corp. | I-Joists and LVLs | 142 |
| Bollen International, Inc. | Paints | 443 |
| Bomanite Corp. | Porous Paving | 17 |
| Bonded Fiber Products, Inc. | Roofing Felt | 309 |
| | Soil Stabilization | 2 |
| Bradford White | Mini Tank Water Heater | 536 |
| Bronx 2000/Big City Forest | Recycled Lumber | 83 |
| Bryant, Day and Night, Payne | Furnace | 551 |
| BTW Industries, Inc. | Site Furnishings | 33 |
| Building Futures | Salvaged Materials | 502 |
| Building Materials Distributors | Salvaged Materials | 503 |
| Building Resources | Salvaged Materials | 504 |
| Burnham Corp. | Boiler | 552 |

## C

| Manufacturer Name | Product Category | Record Number |
|---|---|---|
| C and M Diversified | Salvaged Materials | 505 |
| C F and I Steel LP | Wire and Tubular Steel | 72 |
| Caldwell Building Wrecking | Recycled Lumber | 84 |
| | Salvaged Materials | 506 |
| California Products Corp. | Paints | 444 |
| California Shake Corp. | Fiber-Cement Shingles | 267 |
| Cambridge Designs | Site Furnishings | 34 |
| CAN-CELL Industries, Inc. | Loose-Fill Insulation | 213 |
| Caradco | Wood Windows | 335 |
| Caridon Peachtree Doors and Windows, Inc. | Composite Doors | 322 |
| Carlisle Tire and Rubber Co. | Tile Flooring | 391 |
| Carousel Carpet Mills | Carpet | 417 |
| Carrier Corp. | Furnace | 553 |
| Carrysafe | Plastic Lumber | 173 |
| Cascades Re-Plast, Inc. | Plastic Lumber | 174 |
| Cell-Pak, Inc. | Loose-Fill Insulation | 214 |

| Manufacturer Name | Product Category | Record Number |
|---|---|---|
| Central Fiber Corp. | Loose-Fill Insulation | 215 |
| Centre Mills Antique Wood | Recycled Lumber | 85 |
| CertainTeed Corp. | Roofing Felt | 310 |
| CertainTeed Corp. Insulation Group | Fiberglass Insulation | 216 |
| Challenge Door Co. | Steel Entry Doors | 319 |
| Champagne Industries, Inc. | Vinyl Windows | 336 |
| Champion Insulation | Loose-Fill Insulation | 217 |
| Chemical Specialties, Inc. | Preservative | 161 |
| Cherokee Sanford Brick Co. | Brick Pavers | 15 |
| Chicago Adhesive | Carpet Adhesives | 418 |
| Chicago Metallic | Acoustical Ceilings | 370 |
| Children's Playstructures, Inc. | Playground Equipment | 31 |
| Chronomite Labs | Tankless Water Heater | 537 |
| Classic Products, Inc. | Metal Shingles | 268 |
| Clayville Insulation | Loose-Fill Insulation | 218 |
| Clearwater Tech. | Ozone Purification System | 481 |
| Clivus Multrum, Inc. | Composting Toilet | 529 |
| Colin Campbell Ltd. | Carpet | 419 |
| Collins Pine Co. | Sustainable Harvest | 86 |
| Compak America | Particleboard | 125 |
| Composite Technologies | Composite Fiber Ties | 50 |
| Concrete Designs, Inc. | Ornamental Precast Concrete | 58 |
| Conklin's Authentic Antique Barnwood | Field Stone | 69 |
| | Reclaimed Flooring | 375 |
| | Reclaimed Lumber | 87 |
| Contact Lumber | Trim Products | 151 |
| Controlled Energy Corp. | Mini Tank Water Heater | 539 |
| | Tankless Water Heater | 538 |
| Coronado Paint Co. | Paints | 445 |
| Corrim Co. | Fiberglass Doors | 328 |
| Counter/Productions | Countertop | 513 |
| Coyuchi, Inc. | Textiles | 499 |
| Creative Office Systems, Inc. | Refurbished Office Furniture | 514 |
| Crossroads Recycled Lumber | Recycled Lumber | 89 |
| Crossville Ceramics | Ceramic Tile | 362 |
| Crowe Industries Ltd. | Polymer Shingles | 269 |
| Crown Corp. NA | Wallpaper | 462 |
| Cunningham Brick Co., Inc. | Brick | 60 |
| | Recycled Brick Gravel | 7 |
| Cut and Dried Hardwoods | Sustainable Harvest | 89 |

## D

| Manufacturer Name | Product Category | Record Number |
|---|---|---|
| Dal-Tile Corp. | Ceramic Tile | 363 |
| Davis Colors | Concrete Colorant | 43 |
| Del Industries | Ozone Purification System | 482 |
| Design Materials, Inc. | Wall and Floor Coverings | 462 |
| Design Tex Fabrics | Textiles | 500 |
| Dinoflex Manufacturing Ltd. | Mat Flooring | 392 |
| DLW Gerbert Ltd. | Tile and Sheet Flooring | 393 |
| Dodge-Regupol | Cork Flooring Tiles | 395 |
| | Tile and Sheet Flooring | 394 |
| Duluth Timber Co. | Recycled Lumber | 90 |
| Dura Undercushions, Inc. | Carpet Underlayment | 420 |
| Durable Corp. | Dock Bumper | 480 |
| Durable Mat Co. | Tile and Sheet Flooring | 396 |
| Durox Building Units Ltd. | Aerated Concrete Block | 61 |

| Manufacturer Name | Product Category | Record Number |
|---|---|---|

## E

| | | |
|---|---|---|
| Eagle One Golf Products | Plastic Lumber | 175 |
| | Site Furnishings | 35 |
| | Synthetic Surfacing | 8 |
| Eagle Panel Systems | Manufactured Wall Panels | 284 |
| Eagle Recycled Products | Plastic Lumber | 176 |
| Eagle Windows | Wood Windows | 337 |
| Eaglebrook Products, Inc. | Plastic Lumber | 177 |
| Earth Care Midwest | Site Furnishings | 36 |
| Earth Care Products of America | Plastic Lumber | 178 |
| Earth Safe | Site Furnishings | 37 |
| Earthstar Energy Systems | Water Heat Exchanger | 564 |
| Eco Design/Natural Choice | Cork Flooring Tiles | 397 |
| | Paints and Stains | 446 |
| | Preservative | 162 |
| | Wood Floor Finishes | 376 |
| EcoTimber | Wood Flooring | 377 |
| Edensaw Woods Ltd. | Sustainable Harvest | 91 |
| EEE ZZZ Lay Drain Co., Inc. | Gravel Substitute | 22 |
| EEMAX, Inc. | Tankless Water Heater | 540 |
| EHT Siegmund, Inc. | Radiant Floor Heating and Cooling | 569 |
| Elsro, Inc. | Asphalt Plank Tile | 398 |
| ENER-GRID, Inc. | Polystyrene/Cement Forms | 63 |
| Enercept, Inc. | Manufactured Wall Panels | 285 |
| Energy Pro | Loose-Fill Insulation | 219 |
| Energy Solutions International | Efficient Lighting | 576 |
| Energy Zone Manufacturing, Inc. | Loose-Fill Insulation | 220 |
| Englehard Corp. | Concrete Foaming Liquid | 44 |
| Environmental Plastics, Inc. | Retaining Wall System | 3 |
| Environmental Specialty Products, Inc. | Plastic Lumber | 179 |
| Environmentally Safe Products, Inc. | Foam Insulation | 221 |
| EnviroSafe Products, Inc. | Site Furnishings | 38 |
| Envirotech | Masonry Fireplaces | 468 |
| Envirovac, Inc. | Vacuum Toilet | 530 |
| Eternit | Fiber-Cement Panels | 286 |
| | Fiber-Cement Shingles | 270 |
| Evanite Fiber Corp. | Fiberboard | 152 |
| Exerflex | Plastic Flooring | 436 |

## F

| | | |
|---|---|---|
| Fairmont Corp. | Carpet Underlayment | 421 |
| Feeny Manufacturing Co. | Waste Management System | 479 |
| Fibreworks | Carpet | 422 |
| Fibrex, Inc. | Acoustical Insulation | 371 |
| | Mineral Wool Insulation | 222 |
| Firespaces | Masonry Fireplaces | 469 |
| Flexco Co. | Tile and Sheet Flooring | 399 |
| Flexi-Wall Systems | Plaster Wall Fabric | 463 |
| Flokati Wool Rugs | Area Rugs | 423 |
| Florida Playground and Steel Co. | Playground Equipment | 32 |
| Foam Tech, Inc. | Foam Insulation | 223 |
| Forbo North America | Sheet Flooring | 400 |
| Formica Corp. | Plastic Laminate | 191 |
| | Wood Veneer | 153 |
| Franklin International | Construction Adhesives | 122 |
| Futurebilt, Inc. | Manufactured Wall Panels | 288 |

## G

| Manufacturer Name | Product Category | Record Number |
|---|---|---|
| G. R. Plume Co. | Recycled Lumber | 92 |
| GAF Corp. | Fiber-Cement Shingles | 271 |
| Garland-White and Co. | Thinset Mortar | 59 |
| Gas Purification System | Ozone Purification System | 483 |
| General Electric | Dishwasher | 492 |
| | Refrigerator | 491 |
| | Compact Fluorescent Lamps | 577 |
| Gentek Building Products, Inc. | Vinyl Windows | 338 |
| Georgia-Pacific Corp. | Fiber-Cement Panels | 289 |
| | I-Joists and LVLs | 143 |
| | Organic Shingles | 272 |
| Gerard Roofing Technology | Metal Shingles | 273 |
| Giati Design | Site Furnishings | 39 |
| Gilmer Wood Co. | Sustainable Harvest | 93 |
| Glidden Paint and Wallcovering | Paints | 447 |
| Global Plastic Products, Inc. | Solid Sheet Plastic | 192 |
| Global Technology Systems | Rubber Surfacing | 20 |
| Globe Building Materials, Inc. | Roofing Shingles | 274 |
| Glowcore | Boiler | 554 |
| Goodwin Heart Pine Co. | Reclaimed Flooring | 378 |
| | Recycled Lumber | 94 |
| The Green Paint Co. | Paints | 448 |
| GreenStone Industries, Fort Wayne | Loose-Fill Insulation | 224 |
| GreenStone Industries, Maryland | Loose-Fill Insulation | 225 |
| GreenStone Industries, Norfolk | Loose-Fill Insulation | 226 |
| GreenStone Industries, Sacramento | Loose-Fill Insulation | 227 |
| Greenwood Cotton Insulation | Cotton Insulation | 228 |
| Gridcore Systems | Cellulose Panel | 154 |
| GS Roofing Products Co., Inc. | Roofing Felt | 307 |

## H

| Manufacturer Name | Product Category | Record Number |
|---|---|---|
| Hamilton Manufacturing, Inc. | Loose-Fill Insulation | 229 |
| Handloggers Hardwood | Sustainable Harvest | 95 |
| Harborlite Corp. | Mineral Insulation | 230 |
| Harmony Exchange | Manufactured Wall Panels | 290 |
| Hebel U.S.A. LP | Aerated Concrete Block | 62 |
| Heliotrope General | Ozone Purification System | 484 |
| Hendricksen Naturlich | Carpet | 424 |
| | Carpet Underlayment | 425 |
| Herman Miller, Inc. | Office Furniture | 515 |
| Homasote Co. | Acoustical Insulation | 372 |
| | Carpet Underlayment | 426 |
| | Cellulose-Fiber Decking | 127 |
| | Expansion Joint Filler | 54 |
| | Fiberboard | 155 |
| | Laminated Sheathing | 126 |
| | Manufactured Wall Panels | 291 |
| Honeywell, Inc. | Heat Exchanger Ventilator | 565 |
| Huebert Fiberboard Co. | Board Insulation | 260 |
| Humane Manufacturing Co. | Rubber Walk | 475 |
| Hurd Millworks | Wood Windows | 339 |
| Hydrotherm Corp. | Boiler | 555 |
| Hydrozo, Inc. | Fluid-Applied Water Repellent | 200 |
| Hygrade Glove and Safety | Fences | 30 |

| Manufacturer Name | Product Category | Record Number |
|---|---|---|

## I

| | | |
|---|---|---|
| Icynene, Inc. | Sprayed Insulation | 231 |
| Image Carpets, Inc. | Carpet | 427 |
| In-Cide Technologies | Loose-Fill Insulation | 232 |
| In-Sink-Erator | Mini Tank Water Heater | 541 |
| Indusol, Inc. | Tile Flooring | 401 |
| Innovative Formulation Corp. | Polyester Roofing | 308 |
| Insteel Construction | Polystyrene Concrete Form System | 51 |
| Institute for Sustainable Forestry | Sustainable Harvest | 96 |
| Insul-Tray | Board Insulation | 233 |
| Insulated Masonry Systems, Inc. | Polystyrene/Cement Block | 64 |
| Insulfoam | Expanded Polystyrene | 234 |
| InteGrid Building Systems | Polystyrene/Cement Block | 65 |
| International Cellulose | Sprayed Insulation | 235 |
| International Surfacing, Inc. | Asphaltic Concrete Paving | 9 |
| Into the Woods | Recycled Lumber | 97 |
| Invisible Structures, Inc. | Porous Paving | 18 |

## J

| | | |
|---|---|---|
| J Squared Timberworks, Inc. | Fireplace Mantels | 526 |
| | Recycled Lumber | 98, 156 |
| J-Deck, Inc. Building Systems | Manufactured Wall Panels | 292 |
| Jager Industries, Inc. | I-Joists | 144 |
| James Hardie Building Products | Fiber-Cement Panels | 293 |
| | Fiber-Cement Shingles | 275 |
| Jefferson Recycled Woodworks | Recycled Lumber | 99 |
| Jeld-Wen | Wood Doors | 323 |
| The Joinery Co. | Recycled Flooring | 379 |
| Jotul USA, Inc. | Wood and Gas Stoves | 471 |

## K

| | | |
|---|---|---|
| K and B Associates | Polystyrene Concrete Form System | 52 |
| Kelley-Moore Paint Company | Paints | 449 |
| Keltech, Inc. | Tankless Water Heater | 542 |
| Kent Valley Masonry | Masonry Fireplaces | 471 |
| Kentucky Wood Floors | Parquet Flooring | 380 |
| Key Solutions | Ceramic Insulating Paint | 439 |
| Kilo Alpha Co. | Tankless Water Heater | 543 |
| KitchenAid | Dishwasher | 493 |
| Kohler Co. | Waste Management System | 533 |
| Kolbe and Kolbe Millwork Co. | Wood Windows | 340 |
| KORQ, Inc. | Tile and Sheet Flooring | 402 |
| | Wallpaper | 464 |

## L

| | | |
|---|---|---|
| L. M. Scofield Co. | Concrete Colorant and Sealers | 45 |
| Lamwood Systems, Inc. | Glu-Lam Beam | 133 |
| Lancaster Colony Commercial Products | Tile and Sheet Flooring | 403 |
| Langhorne Carpet Co. | Carpet | 428 |
| Larson Wood Products, Inc. | Sustainable Harvest | 100 |
| Last Chance Mercantile | Salvaged Materials | 507 |
| Lennox Corp. | Boiler | 556 |
| | Furnace | 557 |

| Manufacturer Name | Product Category | Record Number |
|---|---|---|
| Lifeguard Purification Systems | Ozone Purification System | 485 |
| Lincoln Environmental Barrier System | Manufactured Wall Panels | 294 |
| Linford Brothers Glass Co. | Vinyl Windows | 341 |
| Lite-Form, Inc. | Concrete Form System | 53 |
| Loading Dock | Salvaged Materials | 508 |
| Lochinvar Corp. | Boiler | 558 |
| Loewen Windows | Steel Entry Doors | 320 |
| | Wood Windows | 342 |
| Loewenstein | Office Furniture | 516 |
| Lopez Quarries | Masonry Fireplaces | 472 |
| Louisiana-Pacific | Finger-Jointed Studs | 134 |
| | Gypsum Board | 356 |
| | Hardboard Siding | 295 |
| | I-Joists and LVLs | 145 |
| | Laminated Sheathing | 128 |
| | Loose-Fill Insulation | 236 |

## M

| Manufacturer Name | Product Category | Record Number |
|---|---|---|
| M. Susi and Sons | Recycled Concrete Gravel | 10 |
| M. A. Industries, Inc. | Concrete Cylinder Mold | 55 |
| Mameco/Paramount Technical Products | Membrane Waterproofing | 197 |
| Mandish Research International | Landscape Ornaments | 25 |
| Marlite | Fiberboard | 304 |
| Marvin Windows and Door Co. | Composite Window | 343 |
| Masonite Corp. | Hardboard Siding | 296 |
| | Wood Doors | 324 |
| | Wood Shingles | 276 |
| Mat Factory | Tile Flooring | 404 |
| Mats, Inc. | Mats | 522 |
| | Tile and Sheet Flooring | 405 |
| MaxiTile, Inc. | Fiber-Cement Shingles | 277 |
| Maxlite SK America, Inc. | Compact Fluorescent Lamps | 578 |
| Maxwell Pacific | Recycled Lumber | 101 |
| Maytag | Refrigerator | 494 |
| Maze Nails | Nails | 123 |
| Meadowood Industries, Inc. | Particleboard | 157 |
| Medite Corp. | Fiberboard | 158 |
| Menominee Tribal Enterprises | Sustainable Harvest | 102 |
| Merida Meridian, Inc. | Carpet | 429 |
| Metal Sales Manufacturing Corp. | Metal Shingles | 278 |
| Metro Plastics, Inc. | Plastic Lumber | 180 |
| Metropolitan Ceramics | Ceramic Tile | 364 |
| Midwest Faswall, Inc. | Wood/Concrete Forms | 66 |
| Milgard | Vinyl Windows | 344 |
| The Millennium Group | Gypsum Board Clip | 358 |
| Miller Paint Co. | Paints | 450 |
| Miller SQA | Refurbished Office Furniture | 517 |
| Modern Insulation | Loose-Fill Insulation | 237 |
| Mount Storm | Sustainable Harvest | 103 |
| Mountain Fiber Insulation | Loose-Fill Insulation | 238 |
| Mountain Lumber Co. | Recycled Lumber | 104 |
| Multi-Tech Ltd. | Tile and Sheet Flooring | 437 |
| Murco Wall Products, Inc. | Paints | 451 |
| | Plaster/Joint Compound | 357 |
| Myers Architectural | Metal Furniture | 518 |

## N

| | | |
|---|---|---|
| Nascor, Inc. | Manufactered Roofing and Siding | 297 |
| The Natural Bedroom Co. | Bedding and Linens | 519 |
| Natural Cork Ltd. | Tile and Sheet Flooring | 406 |
| Natural Resources | Recycled Lumber | 105 |
| Neenah Foundry Co. | Metal Castings | 78 |
| Nisus Corp. | Preservative | 163 |
| No Fault Industries, Inc. | Tile and Sheet Flooring | 407 |
| Norco Window Co. | Vinyl Windows | 346 |
| | Wood Doors | 325 |
| | Wood Windows | 345 |
| Northern Insulation Products | Loose-Fill Insulation | 239 |
| Nu-Woll Insulation | Loose-Fill Insulation | 240 |
| Nutech | Air Filter | 573 |
| | Heat Exchanger Ventilator | 468 |

## O

| | | |
|---|---|---|
| Obex, Inc. | Plastic Lumber | 183 |
| Off the Wall Architectural Antiques | Salvaged Materials | 509 |
| Old-fashioned Milk Paint Co. | Paints | 452 |
| Omega Salvage | Salvaged Materials | 510 |
| Oregon Strand Board | Floor and Wall Sheathing | 129 |
| Oscoda Plastics, Inc. | Tile and Sheet Flooring | 308 |
| Osram Sylvania | Compact Fluorescent Lamps | 579 |
| Ottawa Fibre, Inc. | Fiberglass Insulation | 241 |
| Owens Corning | Fiberglass Insulation | 242 |
| | Fiberglass Windows | 347 |
| Oxygen Tech. | Ozone Purification System | 486 |
| Ozotech, Inc. | Ozone Purification System | 487 |

## P

| | | |
|---|---|---|
| P. K. Insulation Manufacturing Co. | Loose-Fill Insulation | 243 |
| Pace Chem Industries, Inc. | Paints | 453 |
| Pallas Textiles | Wall Coverings | 465 |
| Palmer Industries | Foam Insulation | 244 |
| | Paints | 454 |
| Panasonic Lighting Co. | Compact Fluorescent Lamps | 580 |
| Paul's Insulation | Loose-Fill Insulation | 245 |
| Pease Industries | Composite Doors | 326 |
| Pella Corp. | Wood Windows | 348 |
| Perma Flake Corp. | Loose-Fill Insulation | 226 |
| Persolite Products, Inc. | Mineral Insulation | 227, 261 |
| Peter Lang Co. | Sustainable Harvest | 106 |
| Phenix Biocomposites, Inc. | Fiberboard | 305 |
| Philips Lighting Co. | Compact Fluorescent Lamps | 581 |
| Phoenix Recycled Plastics | Plastic Lumber | 182 |
| Pioneer Millworks | Recycled Lumber | 107 |
| Pittsford Lumber Co. | Sustainable Harvest | 108 |
| The Plastic Lumber Co. | Plastic Lumber | 183 |
| | Site Furnishings | 40 |
| Plastic Recycling, Inc. | Plastic Lumber | 184 |
| Plastic Tubing, Inc. | Plastic Tubing | 23 |
| Plastipro Canada, Inc. | Tile Flooring | 409 |
| Playfield International | Rubber Pavers | 16 |

| Manufacturer Name | Product Category | Record Number |
|---|---|---|
| | Rubber Surfacing | 21 |
| Premdoor Entry Systems | Fiberglass Doors | 329 |
| Presto Products | Porous Paving | 19 |
| | Retaining Wall System | 4 |
| Prime Line Decorating | Preservative | 164 |
| PrimeBoard, Inc. | Particleboard | 159 |

## Q

| | | |
|---|---|---|
| Quinstone Industries, Inc. | Molded Cellulose Surfaces | 70 |

## R

| | | |
|---|---|---|
| R. C. Musson | Mats | 523 |
| Rasmussen Paint Co. | Paints | 455 |
| RB Rubber Products, Inc. | Mats and Underlayment | 410 |
| | Synthetic Surfacing | 11 |
| RCM International | Tile and Sheet Flooring | 411 |
| Recreation Creations | Site Furnishings | 41 |
| Recycled Plastic Man, Inc. | Plastic Lumber | 185 |
| Recycled Plastics Industries, Inc. | Plastic Lumber | 186 |
| Recycled Polymer Associates | Plastic Lumber | 187 |
| Redco II | Mineral Insulation | 248 |
| Regal Industries, Inc. | Loose-Fill Insulation | 249 |
| Reliance Carpet Cushion | Carpet Underlayment | 430 |
| Renewed Materials Industries | Plastic Lumber | 188 |
| Republic Paints | Paints | 456 |
| Resource Conservation Structures, Inc. | Manufactured Wall Panels | 298 |
| Resource Conservation Tech. | Accessories | 351 |
| | Caulk and Gaskets | 316 |
| | Pond Liner | 24 |
| | Vapor Retarder | 202 |
| Resource Woodworks, Inc. | Recycled Lumber | 109 |
| ReWater Systems, Inc. | Water Conservation System | 488 |
| Rheem/Ruud | Mini Tank Water Heater | 544 |
| | Water Heater | 545 |
| Robert Bosch Corp. | Dishwasher | 495 |
| Robur Corp. | Chiller/Heater | 559 |
| | Refrigerator | 496 |
| Rodman Industries | Particleboard | 381 |
| Rosoboro Lumber Co. | Glu-Lam Beam | 137 |
| Royal Pedic | Bedding and Mattresses | 520 |
| Rubber Polymer Corp. | Fluid-Applied Waterproofing | 198 |

## S

| | | |
|---|---|---|
| Safety Turf | Synthetic Surfacing | 12 |
| Sanger Sales | Salvaged Materials | 511 |
| Santana Plastic Products | Plastic Toilet Partitions | 467 |
| Santana Laminations, Inc. | Solid Sheet Plastic | 193 |
| Sanyo-Fisher USA Corp. | Multizone Ductless Heat Pump | 560 |
| Savnik and Co. Tailors in Wool | Carpet | 431 |
| Savogran | Paint Cleaners | 457 |
| Schuller International, Inc. | Board Insulation | 251 |
| | Fiberglass Insulation | 250 |
| Schuyler Rubber Co. | Walkway and Roadway Additions | 26 |

| Manufacturer Name | Product Category | Record Number |
|---|---|---|
| Scientific Developments, Inc. | Walkway and Roadway Additions | 27 |
| Shelter Enterprises | Manufactured Wall Panels | 299 |
| Sierra Timber Framers | Recycled Lumber | 110 |
| Simplex Products Division | Laminated Sheathing | 130 |
| Slant/Fin Corp. | Boiler | 561 |
| Smith and Fong Co. | Bamboo Flooring | 382 |
| Smith and Hawkins | Site Furnishings | 42 |
| Smurfit Newsprint Corp. | Hardboard Siding | 300 |
| Soil Stabilization Products Co. | Road Surfacing | 13 |
| Solomit Strawboard Pty. Ltd. | Acoustical Ceilings | 373 |
| Southwall Technologies | Insulating Glass | 353 |
| Sparfil Blok Florida, Inc. | Polystyrene/Cement Block | 67 |
| Spartech Plastics | Solid Sheet Plastic | 194 |
| Spectra-Tone Paint Co. | Paints | 458 |
| SPS Marine | Tankless Water Heater | 546 |
| Stadler Corp. | Radiant Floor Hydronic System | 570 |
| Standard Structures, Inc. | I-Joists and Glu-Lams | 138 |
| State Industries | Mini Tank Water Heater | 547 |
| Stein and Collett, Inc. | Recycled Flooring | 383 |
| | Recycled Lumber | 111 |
| Structural Plastics | Plastic Shelving | 476 |
| Summitville Tiles, Inc. | Ceramic Tile | 364 |
| Sun Frost | Refrigerator | 497 |
| Sun Pipe Co., Inc. | Skylight | 313 |
| Suncoast Insulation Manufacturing Co. | Loose-Fill Insulation | 252 |
| Superior Wood Systems | I-Joists | 146 |
| Superlite Block | Concrete Block | 68 |
| Supreme Lighting Co. | Compact Fluorescent Lamps | 582 |
| Sure Fit Shims | Plastic Shims | 124 |
| Surfacing Concepts, Inc. | Rubber Surfacing | 14 |
| Sutherlin Carpet Mills | Carpet | 433 |
| | Carpet Underlayment | 432 |
| Sylan and Dean Brandt | Recycled Flooring | 384 |
| Syndesis Studio | Precast Concrete Surfaces | 71 |
| | Precast Terrazzo | 367 |

## T

| | | |
|---|---|---|
| Tadiran Electrical Appliances | Multizone Ductless Heat Pump | 562 |
| Talisman Mills, Inc. | Carpet | 434 |
| Tallon Termite and Pest Control | Insect Treatment | 165 |
| Tamko Roofing Products, Inc. | Roofing Felt | 311 |
| | Roofing Shingles | 279 |
| Tascon, Inc. | Loose-Fill Insulation | 253 |
| Taylor Building Products | Steel Entry Doors | 321 |
| Tectum, Inc. | Cellulose-Fiber Decking | 131 |
| Teledyne Laars | Boiler | 563 |
| Temp-Cast Enviroheat | Masonry Fireplaces | 473 |
| Temple-Inland Forest Products | Fiberboard | 168 |
| Tenneco Building Products | Board Insulation | 254 |
| Tenneco Packaging/Hexacomb | Honeycomb Cores | 287 |
| Tennessee Cellulose, Inc. | Loose-Fill Insulation | 255 |
| Terra Nativa Sisal | Area Rugs | 435 |
| Terra-Green Technologies | Ceramic Tile | 366 |
| Therma-Stor Products | Heat Exchanger Ventilator | 567 |
| | Heat Pump Water Heater | 548 |
| Therma-Tru Corp. | Composite Doors | 327 |
| ThermaFiber LLC | Mineral Wool Insulation | 256 |

| Manufacturer Name | Product Category | Record Number |
|---|---|---|
| ThermalEase | Radiant Floor Hydronic System | 571 |
| Thermoguard Insulation Co. | Loose-Fill Insulation | 257 |
| 3-10 Insulated Forms LP | Polystyrene Concrete Form System | 46 |
| Timberweld | Glu-Lam Beam | 135 |
| Tosten Brothers Lumber | Sustainable Harvest | 112 |
| Tree Products Hardwoods | Sustainable Harvest | 113 |
| Trex Wood Polymer | Plastic Lumber | 189 |
| Tri-Steel Structures | Steel Studs | 74 |
| Trimax Lumber | Plastic Lumber | 190 |
| Trus Joist MacMillan | Glu-Lam Beam | 139 |
| Trus Joist MacMillan, Paralam Division | Glu-Lam Construction | 136 |
| Truswal | Trusses | 147 |
| Turtle Plastics | Tile and Sheet Flooring | 412 |

## U

| Manufacturer Name | Product Category | Record Number |
|---|---|---|
| U.S. Borax, Inc. | Preservative | 166 |
| U.S. Building Panels, Inc. | Manufactured Wall Panels | 301 |
| UGL | Wood Floor Finishes | 385 |
| Unadilla Laminated Products | Glu-Lams and Columns | 140 |
| Under the Canopy Wood Products | Sustainable Harvest | 114 |
| United McGill Corp. | Duct Mastic | 319 |
| Universal Polymer | Board Insulation | 258 |
| Urban Ore | Salvaged Materials | 511 |
| US-PAR Enterprises | Compact Fluorescent Lamps | 583 |
| USCOA International Corp. | Mats | 524 |

## V

| Manufacturer Name | Product Category | Record Number |
|---|---|---|
| Vaillant Corp. | Tankless Water Heater | 557 |
| Van Duerr Industries | Ramp | 352 |
| Vanport Steel and Supply | Steel Framing | 77 |
| Vent-Aire Systems | Heat Exchanger Ventilator | 568 |
| Vermont Castings, Inc. | Fireplaces and Stoves | 474 |
| Visionwall Technologies, Inc. | Aluminum Windows | 330 |
| Vitra | Office Chairs and Tables | 521 |
| Vulcraft Steel Joist and Girders | Steel Joists and Girders | 6 |

## W

| Manufacturer Name | Product Category | Record Number |
|---|---|---|
| W. F. Taylor Co., Inc. | Floor Adhesives | 413 |
| W. R. Meadows | Concrete Admixtures and Curing | 57 |
|  | Expansion Joint Filler | 56 |
|  | Fluid-Applied Water Repellent | 201 |
| Weather Shield Manufacturing, Inc. | Vinyl Windows | 349 |
| Weatherall Northwest | Preservative | 167 |
| Wellborn Paint | Paints | 459 |
| Wenco Windows | Composite Window | 350 |
| Werzalit of America | Hardboard Siding | 302 |
| Wesco Used Lumber | Recycled Lumber | 115 |
| West Materials, Inc. | Polystyrene Inserts | 259 |
| Weyerhaeuser | Laminated Sheathing | 132 |
| What It's Worth | Recycled Flooring | 386 |
| Whirlpool | Refrigerator | 498 |
| Wild Iris Forestry | Recycled Lumber | 116 |
| Wildwoods Co. | Sustainable Harvest | 117 |
| Willamette Industries Engineered Wood Products | I-Joists, Glu-Lams and Columns | 141 |

| Manufacturer Name | Product Category | Record Number |
|---|---|---|
| William Zinsser and Co. | Paints and Shellacs | 460 |
| Wirsbo Co. | Radiant Floor Hydronic System | 572 |
| Wolverine Technologies | Vinyl Siding | 303 |
| The Wood Cellar | Recycled Flooring | 387 |
| Wood Floors, Inc. | Fireplace Mantels | 527 |
| | Recycled Flooring | 388 |
| | Recycled Lumber | 118 |
| Wooden Workbench | Sustainable Harvest | 119 |
| Woodhouse | Recycled Lumber | 120 |
| Woodworker's Source | Sustainable Harvest | 121 |

## X

| | | |
|---|---|---|
| Xypex Chemical Corp. | Crystalline Barrier Waterproofing | 199 |

## Y

| | | |
|---|---|---|
| Yemm and Hart Green Materials | Plastic Panels | 195 |

## Z

| | | |
|---|---|---|
| Zappone Manufacturing | Metal Shingles | 280 |
| Zwiers, Don and Associates | Fiberglass Shelving | 477 |

# PRODUCT TYPE INDEX

| Product Type | Manufacturer Name | Record Number |
|---|---|---|

## A

| | | |
|---|---|---|
| Accessibility | Van Duerr Industries | 352 |
| Adhesives | AFM Enterprises, Inc. | 196 |
| | Franklin International | 122 |
| | United McGill Corp. | 318 |
| | W. F. Taylor Co., Inc. | 413 |
| Air Purification (see HVAC) | | |
| Appliances, Dishwashers | Asko, Inc. | 490 |
| | General Electric | 492 |
| | KitchenAid | 493 |
| | Robert Bosch Corp. | 496 |
| Appliances, Refrigerator | Amana Refrigeration | 109 |
| | General Electric | 491 |
| | Maytag | 494 |
| | Robur Corp. | 495 |
| | Sun Frost | 497 |
| | Whirlpool | 498 |
| Appliances, Stove | Jotul USA, Inc. | 470 |
| Area Rugs (see Carpet, Area Rugs, and Doormats) | | |

## B

| | | |
|---|---|---|
| Bathroom Fixtures | American Standard | 531 |
| | Aromat Corp. | 528 |
| | BathCrest, Inc. | 438 |
| | Clivus Multrum, Inc. | 529 |
| | Envirovac, Inc. | 530 |
| | Santana Plastic Products | 466 |
| Brick | Cherokee Sanford Brick | 15 |
| | Cunningham Brick Co., Inc. | 60 |
| Bedding | The Natural Bedroom Co. | 519 |
| | Royal Pedic | 520 |

## C

| | | |
|---|---|---|
| Carpet Adhesives | Chicago Adhesive | 418 |
| Carpet, Area Rugs, and Doormats | Allegro Rug Weaving | 414 |
| | Ashland Rubber Mats Co., Inc. | 525 |
| | Bloomsburg Carpet Industries, Inc. | 415 |
| | Blue Ridge Carpet Mills | 416 |
| | Carousel Carpet Mills | 417 |
| | Colin Campbell Ltd. | 419 |
| | Fibreworks | 422 |
| | Flokati Wool Rugs | 423 |
| | Hendricksen Naturlich | 424 |
| | Image Carpets, Inc. | 427 |
| | Langhorne Carpet Co. | 428 |
| | Mats, Inc. | 522 |
| | Merida Meridian, Inc. | 429 |
| | R. C. Musson | 523 |
| | Savnik and Co. Tailors in Wool | 431 |
| | Sutherlin Carpet Mills | 432 |
| | Talisman Mills, Inc. | 434 |
| | Terra Nativa Sisal | 435 |
| | USCOA International Corp. | 524 |
| Carpet Underlayment | Dura Undercushions, Inc. | 420 |

| Product Type | Manufacturer Name | Record Number |
|---|---|---|
| | Fairmont Corp. | 421 |
| | Hendrickson Naturlich | 425 |
| | Homasote Co. | 126 |
| | RB Rubber Products, Inc. | 410 |
| | Reliance Carpet Cushion | 430 |
| | Sutherlin Carpet Mills | 432 |
| Caulk | AFM Enterprises, Inc. | 196 |
| | Auro-Sinan | 315 |
| | Resource Conservation Tech. | 202 |
| Ceilings, Acoustical | Armstrong World Industries | 369 |
| | Chicago Metallic | 370 |
| | Solomit Strawboard Pty., Ltd. | 373 |
| Concrete, Accessories | M. A. Industries, Inc. | 55 |
| Concrete, Mixtures, and Joint Fillers | Davis Colors | 43 |
| | Englehard Corp. | 44 |
| | Homasote Co. | 54 |
| | L. M. Scofield Co. | 45 |
| | Syndesis Studio | 71 |
| | W. R. Meadows | 56, 201 |
| Concrete Block | Durox Building Units | 61 |
| | ENER-GRID, Inc. | 63 |
| | Hebel U.S.A. | 62 |
| | Insulated Masonry Systems, Inc. | 64 |
| | InteGrid Building Systems | 65 |
| | Midwest Faswall, Inc. | 66 |
| | Sparfil Blok Florida, Inc. | 67 |
| | Superlite Block | 68 |
| Concrete, Ornamental | Concrete Designs, Inc. | 58 |
| | Mandish Research International | 25 |
| Countertops | Counter/Productions | 513 |
| | Phenix Biocomposites, Inc. | 305 |
| | Quinstone Industries, Inc. | 70 |
| | Syndesis Studio | 71 |

## D

| | | |
|---|---|---|
| Decking | Homasote Co. | 127 |
| | Tectum, Inc. | 131 |
| Doors | Advanced Environmental Recycling Technologies | 148 |
| | Caridon Peachtree Doors and Windows, Inc. | 322 |
| | Challenge Door Co. | 319 |
| | Corrim Co. | 328 |
| | Jeld-Wen | 323 |
| | Loewen Windows | 320 |
| | Masonite Corp. | 324 |
| | Norco Window Co. | 325 |
| | Pease Industries | 326 |
| | Premdoor Entry Systems | 329 |
| | Taylor Building Products | 321 |
| | Therma-Tru Corp. | 327 |

## E

| | | |
|---|---|---|
| Electrical Accessories | Resource Conservation Tech. | 202 |

## F

| | | |
| --- | --- | --- |
| Fabric (see Textiles) | | |
| Fasteners, Nails | C F and I Steel LP | 72 |
| | Maze Nails | 123 |
| Fasteners, Wall Board Clips | The Millennium Group | 358 |
| Fencing | All Fiberglass Products Corp. | 29 |
| | Eagle One Golf Products | 35 |
| | Earth Safe | 37 |
| | Hygrade Glove and Safety | 30 |
| Fireplaces | Biofire, Inc. | 467 |
| | Envirotech | 468 |
| | Firespaces | 469 |
| | Kent Valley Masonry | 471 |
| | Lopez Quarries | 472 |
| | Quinstone Industries, Inc. | 70 |
| | Temp-Cast Enviroheat | 473 |
| | Vermont Castings, Inc. | 474 |
| Fireproofing | American Sprayed Fibers, Inc. | 262 |
| | Tascon, Inc. | 253 |
| Flooring | Albany Woodworks | 374 |
| | Alpha Granite and Marble | 389 |
| | Amtico International | 390 |
| | Bedrock Industries | 360, 361 |
| | Carlisle Tire and Rubber Co. | 391 |
| | Conklin's Authentic Antique Barnwood | 69, 375 |
| | Dinoflex Manufacturing | 392 |
| | DLW Gerbert Ltd. | 393 |
| | Dodge-Regupol | 394, 395 |
| | Durable Mat Co. | 396 |
| | Eco Design/Natural Choice | 397 |
| | Eco Timber | 377 |
| | Elsro, Inc. | 398 |
| | Exerflex | 436 |
| | Flexco Co. | 399 |
| | Forbo North America | 400 |
| | Goodwin Heart Pine Co. | 378 |
| | Indusol, Inc. | 401 |
| | The Joinery Co. | 379 |
| | Kentucky Wood Floors | 380 |
| | KORQ, Inc. | 402 |
| | Lancaster Colony Commercial Products | 403 |
| | Mat Factory | 404 |
| | Mats, Inc. | 405 |
| | Multi-Tech, Ltd. | 437 |
| | Natural Cork, Ltd. | 406 |
| | No Fault Industries, Inc. | 407 |
| | Oscoda Plastics, Inc. | 408 |
| | Phenix Biocomposites, Inc. | 305 |
| | Plastipro Canada, Inc. | 409 |
| | RCM International | 411 |
| | Smith and Fong Co. | 382 |
| | Stein and Collett, Inc. | 383 |
| | Sylan and Dean Brandt | 384 |
| | Syndesis Studio | 71 |
| | Turtle Plastics | 412 |
| | What It's Worth | 386 |

| Product Type | Manufacturer Name | Record Number |
|---|---|---|
| | The Wood Cellar | 387 |
| | Wood Floors, Inc. | 388 |
| Floor Finishes | Eco Design/Natural Choice | 376 |
| | UGL | 385 |
| Forms/Foundations | 3-10 Insulated Forms LP | 46 |
| (see also Concrete Block) | Advanced Foam Plastics, Inc. | 47 |
| | American ConForm Industries, Inc. | 48 |
| | American Polysteel Forms | 49 |
| | Composite Technologies | 50 |
| | ENER-GRID, Inc. | 63 |
| | Insteel Construction | 51 |
| | K and B Associates | 52 |
| | Lite-Form Inc. | 53 |
| | Midwest Faswall, Inc. | 66 |
| | West Materials, Inc. | 259 |
| Framing Lumber (see Lumber) | | |
| Furniture, Office | Creative Office Systems, Inc. | 514 |
| | Herman Miller, Inc. | 515 |
| | Lowenstein | 516 |
| | Miller SQA | 517 |
| | Myers Architectural | 518 |
| | Vitra | 521 |
| Furniture, Outdoor | BTW Industries, Inc. | 33 |
| | Cambridge Designs | 34 |
| | Eagle One Golf Products | 35 |
| | Earth Care Midwest | 36 |
| | Earth Safe | 37 |
| | EnviroSafe Products, Inc. | 38 |
| | Giati Design | 39 |
| | The Plastic Lumber Co. | 40 |
| | Recreation Creations | 41 |
| | Smith and Hawkins | 42 |

## G

| | | |
|---|---|---|
| Gravel | Cunningham Brick Co. Inc. | 7 |
| | EEE ZZZ Lay Drain Co., Inc. | 22 |
| | M. Susi and Sons | 10 |
| Gutters | Bass and Hays Foundry, Inc. | 312 |

## H

| | | |
|---|---|---|
| Heating (see HVAC) | | |
| HVAC, Air Filters | Nutech | 573 |
| HVAC, Boilers | Burnham Corp. | 552 |
| | Glowcore | 554 |
| | Hydrotherm Corp. | 555 |
| | Lennox Corp. | 556 |
| | Lochinvar Corp. | 558 |
| | Slant/Fin Corp. | 561 |
| | Teledyne Laars | 563 |
| HVAC, Chillers/Heaters | Robur Corp. | 559 |
| HVAC, Furnaces | American Standard | 550 |
| | Bryant, Day and Night, Payne | 551 |
| | Carrier Corp. | 553 |
| | Lennox Corp. | 557 |

| Product Type | Manufacturer Name | Record Number |
|---|---|---|
| HVAC, Heat Exchanger Ventilator | Honeywell, Inc. | 565 |
| | Nutech | 566 |
| | Therma-Stor Products | 567 |
| | Vent-Aire Systems | 568 |
| HVAC, Heat Pumps | Sanyo-Fisher USA Corp. | 560 |
| | Tadiran Electrical Appliances | 562 |
| HVAC, Ozone Purification | Aquazone Products Co. | 480 |
| | Clearwater Tech. | 481 |
| | Del Industries | 482 |
| | Gas Purification Systems | 483 |
| | Heliotrope General | 484 |
| | Lifeguard Purification Systems | 485 |
| | Oxygen Tech. | 486 |
| | Ozotech, Inc. | 487 |
| HVAC, Radiant Heating | EHT Sigmund, Inc. | 569 |
| | Stadler Corp. | 570 |
| | ThermalEase | 571 |
| | Wirsbo Co. | 572 |
| HVAC, Water Heaters | Advanced Tech Industries | 534 |
| | American Water Heater | 535 |
| | Bradford White | 536 |
| | Chronomite Labs | 537 |
| | Controlled Energy Corp. | 538, 539 |
| | Earthstar Energy Systems | 564 |
| | EEMAX, Inc. | 540 |
| | In-Sink-Erator | 541 |
| | Keltech, Inc. | 542 |
| | Kilo Alpha Co. | 543 |
| | Rheem/Ruud | 544, 545 |
| | SPS Marine | 546 |
| | State Industries | 547 |
| | Therma-Stor Products | 548 |
| | Vaillant Corp. | 549 |

## I

| Product Type | Manufacturer Name | Record Number |
|---|---|---|
| Insulation, Acoustical | American Sprayed Fibers, Inc. | 207 |
| | Fibrex, Inc. | 222 |
| | Homasote Co. | 155 |
| Insulation, Board | BMCA Insulation Products | 212 |
| | Huebert Fiberboard Co. | 260 |
| | Schuller International | 250 |
| | Tenneco Building Products | 254 |
| | Universal Polymer | 258 |
| Insulation, Cotton | Greenwood Cotton Insulation | 228 |
| Insulation, Fiberglass | CertainTeed Corp. Insulation Group | 216 |
| | Ottawa Fibre, Inc. | 242 |
| | Owens Corning | 243 |
| | Schuller International | 251 |
| Insulation, Foam | American Sprayed Fibers, Inc. | 207 |
| | Environmentally Safe Products, Inc. | 221 |
| | Foam Tech, Inc. | 223 |
| | Icynene, Inc. | 231 |
| | Palmer Industries | 244 |
| Insulation, Loose-Fill | All-Weather Insulation Co. | 204 |
| | American Insulation, Inc. | 205 |
| | Applegate Manufacturing | 208, 209 |
| | Arctic Manufacturing | 210 |

| Product Type | Manufacturer Name | Record Number |
|---|---|---|
| | CAN-CELL Industries, Inc. | 213 |
| | Cell-Pak, Inc. | 214 |
| | Central Fiber Corp. | 215 |
| | Champion Insulation | 217 |
| | Clayville Insulation | 218 |
| | Energy Pro | 219 |
| | Energy Zone Manufacturing, Inc. | 220 |
| | Greenstone Industries | 224, 227 |
| | Hamilton Manufacturing, Inc. | 229 |
| | In-Cide Technologies | 232 |
| | Insul-Tray | 233 |
| | Louisiana-Pacific | 236 |
| | Modern Insulation | 237 |
| | Mountain Fiber Insulation | 238 |
| | Northern Insulation Products | 239 |
| | Nu-Woll Insulation | 240 |
| | P. K. Insulation Manufacturing Co. | 243 |
| | Paul's Insulation | 245 |
| | Perma-Flake Corp. | 246 |
| | Regal Industries, Inc. | 249 |
| | Suncoast Insulation Manufacturing | 252 |
| | Tascon, Inc. | 253 |
| | Tennessee Cellulose, Inc. | 255 |
| | Thermoguard Insulation Co. | 257 |
| Insulation, Mineral | Harborlite Corp. | 230 |
| | Persolite Products, Inc. | 247, 261 |
| | Redco II | 248 |
| Insulation, Mineral Wool | American Rockwool, Inc. | 206 |
| | Fibrex, Inc. | 221 |
| | ThermaFiber, LLC | 256 |
| Insulation, Polystyrene | Advanced Foam Plastics, Inc. | 202 |
| | Insulfoam | 234 |
| | West Materials, Inc. | 259 |
| Insulation, Sprayed | American Sprayed Fibers, Inc. | 207, 262 |
| | Ark-Seal, Inc. International | 211 |
| | Icynene, Inc. | 231 |
| | International Cellulose | 235 |

## L

| | | |
|---|---|---|
| Landscaping | AquaPore Moisture Systems | 28 |
| | Green Systems, Inc. | 1 |
| | Bonded Fiber Products, Inc. | 2 |
| | Mandish Research International | 25 |
| | Neenah Foundry Co. | 78 |
| | Plastic Tubing, Inc. | 23 |
| | Resource Conservation Tech. | 24 |
| | ReWater Systems, Inc. | 488 |
| Lighting | Aero-Tech Lightbulb Co. | 574 |
| | Beacon Light Products, Inc. | 575 |
| | Energy Solutions | 576 |
| | General Electric Co. | 577 |
| | Maxlite SK America | 578 |
| | Osram Sylvania | 579 |
| | Panasonic Lighting Co. | 580 |
| | Philips Lighting Co. | 581 |
| | Supreme Lighting Co. | 582 |
| | US-PAR Enterprises | 583 |

| Product Type | Manufacturer Name | Record Number |
|---|---|---|
| Lumber, Engineered | Boise Cascade Corp. | 142 |
| | Georgia-Pacific Corp. | 143 |
| | Jager Industries, Inc. | 144 |
| | Lamwood Systems, Inc. | 133 |
| | Louisiana-Pacific | 134, 145 |
| | Rosoboro Lumber | 137 |
| | Standard Structures, Inc. | 138 |
| | Superior Wood Systems | 146 |
| | Timberweld | 135 |
| | Trus Joist MacMillan | 136, 139 |
| | Unadilla Laminated Products | 140 |
| | Willamette Industries Engineered Wood Products | 141 |
| Lumber, Plastic | Advanced Environmental Recycling Technologies | 148 |
| | Aeolian Enterprises | 169 |
| | Alket Industries | 170 |
| | Amazing Recycled Products | 171 |
| | Bedford Industries | 172 |
| | Carrysafe | 173 |
| | Cascades Re-Plast, Inc. | 174 |
| | Eagle One Golf Products | 8, 175 |
| | Eagle Recycled Products | 176 |
| | Eaglebrook Products, Inc. | 177 |
| | Earth Care Products of America | 178 |
| | Environmental Specialty Products, Inc. | 179 |
| | Metro Plastics, Inc. | 180 |
| | Obex, Inc. | 181 |
| | Phoenix Recycled Plastics | 182 |
| | The Plastic Lumber Co. | 183 |
| | Plastic Recycling, Inc. | 184 |
| | Recycled Plastic Man, Inc. | 185 |
| | Recycled Plastics Industries | 186 |
| | Recycled Polymer Associates | 187 |
| | Renewed Materials Industries | 188 |
| | Sure Fit Shims | 124 |
| | Trex Wood Polymer | 189 |
| | Trimax Lumber | 190 |
| Lumber, Reclaimed and Recycled (see also Salvaged Materials) | Aged Woods, Inc. | 79 |
| | Allied Demolition, Inc. | 80 |
| | Bronx 2000/Big City Forest | 83 |
| | Caldwell Building Wrecking | 84 |
| | Centre Mills Antique Wood | 85 |
| | Conklin's Authentic Antique Barnboard | 87 |
| | Crossroads Recycled Lumber | 88 |
| | Duluth Timber Co. | 90 |
| | Goodwin Heart Pine Co. | 94 |
| | G. R. Plume Co. | 92 |
| | Into the Woods | 97 |
| | J Squared Timberworks, Inc. | 98, 156 |
| | Jefferson Recycled Woodworks | 99 |
| | Maxwell Pacific | 101 |
| | Mountain Lumber Co. | 104 |
| | Natural Resources | 105 |
| | Pioneer Millworks | 107 |
| | Resource Woodworks, Inc. | 109 |
| | Sierra Timber Framers | 110 |
| | Stein and Collett, Inc. | 111 |
| | Wesco Used Lumber | 115 |

| | | |
| --- | --- | --- |
| | Wild Iris Forestry | 116 |
| | Wood Floors, Inc. | 118 |
| | Woodhouse | 120 |
| Lumber, Sustainable Harvest | Bear Creek Lumber Co. | 81 |
| | Big Creek Lumber Co. | 82 |
| | Collins Pine Co. | 86 |
| | Cut and Dried Hardwoods | 89 |
| | Edensaw Woods Ltd. | 91 |
| | Gilmer Wood Co. | 93 |
| | Handloggers Hardwood | 95 |
| | Institute for Sustainable Forestry | 96 |
| | Larson Wood Products, Inc. | 100 |
| | Menominee Tribal Enterprises | 102 |
| | Mount Storm | 103 |
| | Peter Lang Co. | 106 |
| | Pittsford Lumber Co. | 108 |
| | Tosten Brothers Lumber | 112 |
| | Tree Products Hardwoods | 113 |
| | Under the Canopy Wood Products | 114 |
| | Wildwoods Co. | 117 |
| | Wooden Workbench | 119 |
| | Woodworker's Source | 121 |

## M

| | | |
| --- | --- | --- |
| Masonry | Conklin's Authentic Antique Barnwood | 69 |
| | Quinstone Industries, Inc. | 70 |
| Mortar, Thinset | Garland-White and Co. | 59 |

## P

| | | |
| --- | --- | --- |
| Paint and Stain | AFM Enterprises, Inc. | 196, 440 |
| | Auro-Sinan Co. | 441 |
| | Benjamin Moore and Co. | 442 |
| | Bollen International, Inc. | 443 |
| | California Products Corp. | 444 |
| | Coronado Paint Co. | 445 |
| | Eco Design/Natural Choice | 162 |
| | Glidden Paint and Wallcovering | 447 |
| | The Green Paint Co. | 448 |
| | Kelley-Moore Paint Company | 449 |
| | Miller Paint Co. | 450 |
| | Murco Wall Products, Inc. | 451 |
| | Old-Fashioned Milk Paint Co. | 452 |
| | Pace Chem Industries, Inc. | 453 |
| | Palmer Industries | 454 |
| | Rasmussen Paint Co. | 455 |
| | Republic Paints | 456 |
| | Spectra-Tone Paint Co. | 458 |
| | Wellborn Paint | 459 |
| | William Zinsser and Co. | 460 |
| Particleboard | BioFab | 150 |
| | Compak America | 125 |
| | Meadowood Industries, Inc. | 157 |
| | PrimeBoard Inc. | 159 |
| | Rodman Industries | 381 |
| Pavers | Alpha Granite and Marble | 389 |

| Product Type | Manufacturer Name | Record Number |
|---|---|---|
| | Carlisle Tire and Rubber Co. | 391 |
| | Cherokee Sanford Brick | 15 |
| | Cunningham Brick Co., Inc. | 7 |
| | Dinoflex Manufacturing Ltd. | 393 |
| | Dodge-Regupol | 394 |
| | Flexco Co. | 399 |
| | Mat Factory | 404 |
| | Mats, Inc. | 405 |
| | Playfield International | 16 |
| | RCM International | 411 |
| | Turtle Plastics | 412 |
| Paving | AMREC | 6 |
| | Bomanite Corp. | 17 |
| | International Surfacing, Inc. | 9 |
| | Invisible Structures, Inc. | 18 |
| | Presto Products | 19 |
| | Soil Stabilization Products Co. | 13 |
| Pest Control | Tallon Termite and Pest Control | 165 |
| Plaster and Joint Compound | Auro-Sinan Co. | 354 |
| | Murco Wall Products, Inc. | 357 |
| Plastic Laminate | Formica Corp. | 191 |
| Playground Equipment | Children's Playstructures, Inc. | 31 |
| | Florida Playground and Steel Co. | 32 |
| Plumbing | Blanco America | 532 |
| | Clivus Multrum | 529 |
| | Feeney Manufacturing Co. | 479 |
| | Kohler Co. | 533 |
| | ReWater Systems, Inc. | 488 |
| Preservative and Treatment, Wood | Chemical Specialties, Inc. | 161 |
| | Eco Design/Natural Choice | 162 |
| | Nisus Corp. | 163 |
| | Prime Line Decorating | 164 |
| | U.S. Borax, Inc. | 166 |
| | Tallon Termite and Pest Control | 165 |
| | Weatherall Northwest | 167 |

## R

| Product Type | Manufacturer Name | Record Number |
|---|---|---|
| Radiant Heating (see HVAC) | | |
| Retaining Walls | Bestmann Green Systems, Inc. | 1 |
| | Environmental Plastics, Inc. | 3 |
| | Presto Products | 4 |
| Roof Trusses | Truswal | 147 |
| Roofing | Atlas Roofing Corp. | 266 |
| | Bedard Cascades, Inc. | 306 |
| | Bonded Fiber Products, Inc. | 309 |
| | CertainTeed Corp. | 310 |
| | Crowe Industries, Ltd. | 249 |
| | Georgia-Pacific Corp. | 143, 272 |
| | Globe Building Materials, Inc. | 276 |
| | GS Roofing Products Co., Inc. | 307 |
| | Innovative Formulation Corp. | 308 |
| | Masonite Corp. | 276, 296 |
| | Nascor, Inc. | 297 |
| | Schuller International, Inc. | 251 |
| | Tamko Roofing Products, Inc. | 279 |
| Roofing, Fiber Cement Shingles | American Cemwood Products | 263 |

| Product Type | Manufacturer Name | Record Number |
|---|---|---|
| | California Shake Corp. | 267 |
| | Eternit | 270 |
| | GAF Corp. | 271 |
| | James Hardie Building Products | 275 |
| | MaxiTile, Inc. | 277 |
| Roofing, Metal | ATAS International, Inc. | 264, 265 |
| | Classic Products, Inc. | 268 |
| | Gerard Roofing Technology | 273 |
| | Metal Sales Manufacturing Corp. | 278 |
| | Zappone Manufacturing | 280 |
| Rubber Matting | Dinoflex Manufacturing Ltd. | 392 |
| | Durable Corp. | 478 |
| | Mats, Inc. | 522 |
| | R. C. Musson | 523 |
| | USCOA International Corp. | 524 |
| Rubber Surfacing | Eagle One Golf Products | 8 |
| | Global Technology Systems | 20 |
| | Humane Manufacturing Co. | 475 |
| | No-Fault Industries, Inc. | 407 |
| | Playfield International | 21 |
| | RB Rubber Products, Inc. | 11, 410 |
| | Safety Turf | 12 |
| | Schuyler Rubber Co. | 26 |
| | Scientific Developments, Inc. | 27 |
| | Surfacing Concepts, Inc. | 14 |

## S

| Product Type | Manufacturer Name | Record Number |
|---|---|---|
| Salvaged Materials | Berkeley Architectural Salvage | 501 |
| | Building Futures | 502 |
| | Building Materials Distributors | 503 |
| | Building Resources | 504 |
| | C and M Diversified | 505 |
| | Caldwell Building Wrecking | 506 |
| | Last Chance Mercantile | 507 |
| | Loading Dock | 508 |
| | Off the Wall Architectural Antiques | 509 |
| | Omega Salvage | 510 |
| | Sanger Sales | 511 |
| | Urban Ore | 512 |
| Sheathing | Barrier Technology, USA | 160 |
| | Homasote Co. | 126 |
| | Louisiana-Pacific | 128 |
| | Oregon Strand Board | 129 |
| | Simplex Products Division | 130 |
| | Weyerhaeuser | 132 |
| Shelving | Creative Office Systems, Inc. | 514 |
| | Myers Architectural | 518 |
| | Structural Plastics | 476 |
| | Zwiers, Don and Associates | 477 |
| Siding | Louisiana-Pacific | 128, 295 |
| | Masonite Corp. | 296 |
| | Nascor, Inc. | 297 |
| | Smurfit Newsprint Corp. | 300 |
| | Werzalit of America | 302 |
| | Wolverine Technologies | 303 |
| Skylight | Sun Pipe Co., Inc. | 313 |

| Product Type | Manufacturer Name | Record Number |
|---|---|---|
| Solvents | AFM Enterprises, Inc. | 440 |
| | Auro-Sinan Co. | 441 |
| | Eco Design/Natural Choice | 446 |
| | Savogran | 457 |
| Steel Framing | Advanced Framing Systems, Inc. | 76 |
| | Angeles Metal Systems | 73 |
| | Tri-Steel Structures | 74 |
| | Vanport Steel and Supply | 77 |
| | Vulcraft Steel Joist and Girders | 75 |

## T

| Product Type | Manufacturer Name | Record Number |
|---|---|---|
| Textiles | Coyuchi, Inc. | 499 |
| | Design Tex Fabrics | 500 |
| Tile | Bedrock Industries | 361 |
| | Crossville Ceramics | 362 |
| | Dal-Tile Corp. | 363 |
| | Metropolitan Ceramics | 364 |
| | Summitville Tiles, Inc. | 365 |
| | Terra-Green Technologies | 366 |
| Traffic Control | Amazing Recycled Products | 5 |
| | BTW Industries, Inc. | 33 |
| | Durable Corp. | 480 |
| | Florida Playground and Steel Co. | 32 |
| | Schuyler Rubber Co. | 26 |
| | Scientific Developments, Inc. | 27 |
| Trim | Contact Lumber | 151 |
| | Georgia-Pacific Corp. | 289 |
| | J Squared Timberworks, Inc. | 98, 156 |
| | Wood Floors, Inc. | 118 |

## W

| Product Type | Manufacturer Name | Record Number |
|---|---|---|
| Wall Panels, Fiber-Cement | Eternit | 286 |
| | James Hardie Building Products | 293 |
| Wall Panels, Fiberboard | Architectural Forest Enterprises | 149 |
| | Evanite Fiber Corporation | 152 |
| | Homasote Co. | 126, 155 |
| | Marlite | 304 |
| | Medite Corp. | 158 |
| | Phenix Biocomposites, Inc. | 305 |
| | Temple-Inland Forest Products | 168 |
| Wall Panels, Gypsum Board | Louisiana-Pacific | 356 |
| Wall Panels, Manufactured | AFM Corp. | 281 |
| | Agriboard Industries | 282 |
| | Bellcomb Technologies | 283 |
| | Eagle Panel Systems | 284 |
| | Enercept, Inc. | 285 |
| | Futurebilt, Inc. | 288 |
| | Harmony Exchange | 290 |
| | Homasote Co. | 126, 291 |
| | J-Deck, Inc. Building Systems | 292 |
| | Lincoln Environmental Barrier System | 294 |
| | Resource Conservation Structures, Inc. | 298 |
| | Sheltar Enterprises | 299 |
| | U.S. Building Panels, Inc. | 301 |
| Wall Panels, Paper | Gridcore Systems | 154 |
| | Tenneco Packaging/Hexacomb | 287 |

| Product Type | Manufacturer Name | Record Number |
|---|---|---|
| Wall Panels, Plastic | Global Plastic Products, Inc. | 192 |
| | Santana Plastic Products | 467 |
| | Yemm and Hart Green Materials | 195 |
| Wallcovering | Crown Corp. NA | 461 |
| | Design Materials, Inc. | 462 |
| | Flexi-Wall Systems | 463 |
| | KORQ, Inc. | 464 |
| | Pallas Textiles | 465 |
| Water Heaters (see HVAC, Water Heaters) | | |
| Waterproofing and Water Repellents | AFM Enterprises, Inc. | 196 |
| | Global Plastic Products, Inc. | 192 |
| | Hydrozo, Inc. | 200 |
| | Mameco/Paramount Technical Products | 197 |
| | Resource Conservation Tech. | 24 |
| | Rubber Polymer Corp. | 198 |
| | Spartech Plastics | 194 |
| | W. R. Meadows | 57, 201 |
| | Xypex Chemical Corporation | 199 |
| Windows and Glass | Advanced Environmental Recycling Technologies | 148 |
| | Alaska Window Co. | 331 |
| | Alenco | 332 |
| | Alside | 333 |
| | Andersen Windows | 334 |
| | Caradco | 335 |
| | Champagne Industries, Inc. | 336 |
| | Eagle Windows | 337 |
| | Gentek Building Products, Inc. | 338 |
| | Hurd Millworks | 339 |
| | Kolbe and Kolbe Millwork Co. | 340 |
| | Linford Brothers Glass Co. | 341 |
| | Loewen Windows | 342 |
| | Marvin Window and Door Co. | 343 |
| | Milgard | 344 |
| | Norco Window Co. | 345, 346 |
| | Owens Corning | 347 |
| | Pella Corporation | 348 |
| | Southwall Technologies | 353 |
| | Visionwall Technologies | 330 |
| | Weather Shield Manufacturing, Inc. | 349 |
| | Wenco Windows | 350 |
| Wire | C F and I Steel LP | 72 |
| Wood Stoves (see Fireplaces) | | |
| Wood Veneer | Formica Corp. | 153 |